Introduction to Molecular Motion in Polymers

Introduction to Molecular Motion in Polymers

Professor Richard A. Pethrick
Research Professor, Department of Pure and Applied Chemistry at University of Strathclyde, UK
BSc, PhD, DSc, CChem, FRSC, FRSE, FIM

Associate Professor Taweechai Amornsakchai
Associate Professor, Department of Chemistry at Mahidol University, Thailand
BSc, PhD

Professor Alastair M. North OBE
Visiting Professor, Department of Chemistry at Mahidol University, Thailand
BSc, PhD, DSc, FRSE, FRSC

Whittles Publishing

Published by
Whittles Publishing,
Dunbeath,
Caithness KW6 6EY,
Scotland, UK

www.whittlespublishing.com

Distributed in North America by
CRC Press LLC,
Taylor and Francis Group,
6000 Broken Sound Parkway NW, Suite 300,
Boca Raton, FL 33487, USA

ISBN 978184995-008-4
USA ISBN 978-1-4398-6603-0

The publisher and authors have used their best efforts in preparing this book, but assume
no responsibility for any injury and/or damage to persons or property from the use or
implementation of any methods, instructions, ideas or materials contained within this book.
All operations should be undertaken in accordance with existing legislation and recognized
trade practice. Whilst the information and advice in this book is believed to be true and
accurate at the time of going to press, the authors and publisher accept no legal responsibility
or liability for errors or omissions that may have been made.

Typeset by
iPLUS Knowledge Solutions Private Limited, Chennai-32, India.

Printed by Martins the Printers, Berwick upon Tweed

Contents

Preface

Since the publication in 1981 of our text on molecular motion in polymers (R. T. Bailey, A. M. North and R. A. Pethrick, *Molecular Motion in High Polymers*, Oxford University Press, Oxford, 1981) it has become apparent that there is widespread ignorance, in universities and in persons working with plastics and rubbers, of many aspects of the molecular behaviour underlying the important technological properties of these materials. In particular, fundamental time- and temperature-dependences of the physical properties of plastics and rubbers are poorly understood. Several books on polymer physical properties for specialist research workers have appeared, but in general these contain physics written by physicists for physics students, chemistry written by chemists for chemistry students, or general materials science written by materials scientists for students of materials science. Some, like our text mentioned above, contain too much mathematics for the average reader, while others fail adequately to relate molecular processes to physical phenomena. What is missing is a book giving an easy to understand introduction to an interdisciplinary field, between chemistry, physics, and technology, which might be called "applied polymer molecular physics".

This small volume, drawn in part from our experience of teaching the subject, is an attempt to meet this deficiency.

1

Introduction

Polymers, as plastics and rubbers, pervade our lives and we come across them in many different forms. As such, their physical properties have great importance, and an understanding of them is vital for their uses in technology and engineering.

Polymers possess material properties which are distinctly different from those exhibited by metals, ceramics and glasses. Metals on heating can be transformed from hard solids to low viscosity liquids over a relatively small temperature range. Ceramics exhibit a hardness that does not vary significantly with temperature up to the melting point and have poor impact properties and low elasticity. In contrast, polymers can be hard at low temperatures, comparable to metals or glasses, but on heating can be transformed into a rubbery state and exhibit a high degree of elasticity. Increasing the temperature further in certain cases allows the polymer to be converted into a free flowing liquid. However, some polymers do not flow when heated and so cannot be reshaped.

Although the fact is often ignored, the properties are a direct consequence of the chemical structure of the polymer molecules. So in this text we shall introduce the ways in which the chemical structure determines the physical properties, paying particular attention to how this is affected by temperature and time.

1.1 The bridges of understanding

The objective of this book is to build "bridges of understanding" between, on one hand, the chemical structure of polymer molecules and, on the other, the macroscopic technological or engineering behaviour of a plastic or rubber material. We have opted to do this by way of consideration of molecular behaviour, and particularly of the molecular movement and the morphology of the material. By understanding how chemical structure determines the characteristics of molecular motion and morphology, and how these in turn influence the bulk properties, we can link chemistry to materials science in a smooth and informative way.

CHEMISTRY **TECHNOLOGY AND ENGINEERING**

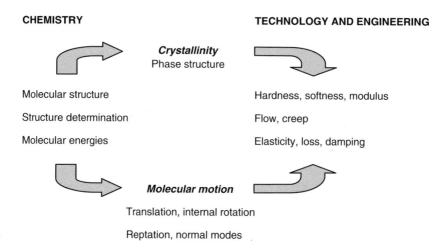

Crystallinity
Phase structure

Molecular structure   Hardness, softness, modulus

Structure determination   Flow, creep

Molecular energies   Elasticity, loss, damping

Molecular motion

Translation, internal rotation

Reptation, normal modes

Figure 1.1 *The two bridges of understanding for polymer science.*

There are two inter-related ways to build this link. We start by asking, "What are the molecules doing?" The answer, of course, is, "The molecules are moving (or not moving)." So we follow that up by asking, "Why are they moving? How are they moving? What controls the movement? What are the consequences of the movement?"

The second way to construct this link is to consider how the structure of the molecules gives rise to crystallinity or other two-phase behaviour, then to examine how the dynamic properties of the morphology link to the physical behaviour. The phase behaviour is a reflection of the balance between the inter- and intra-atomic interactions which determine how easily molecules pack together and also how easily they are able to change their shape.

Of course the two lines of thought are interconnected. To gain a complete insight into how chemical structure predetermines the important technological physical properties we have to consider both processes together. So the third step is to link the morphology to the molecular dynamics. This philosophy is summarised in Figure 1.1.

So in this volume we shall start by examining the chemical molecular structure of plastics and rubbers, and then the shapes of the molecules and how these change under the constraints of temperature, time and available space. At this point we shall consider how the chemical structure influences the bulk morphology, and how this in turn affects the molecular motion. With an understanding of the dynamics of molecular movement and the morphology, we go on to examine the ways that these influence the bulk mechanical properties.

The mechanical properties are the principal reason for the technical importance of polymer materials, but they are not the only one. After the section on mechanical properties, we shall investigate the electrical properties, then the photo properties and finally some diffusion properties, in each case relating the technological aspects to the underlying molecular dynamics.

Further reading

Grosberg Yu. and Khokhlov A.R. *Giant Molecules*, Academic Press, San Diego, 1997.

2

Chemical structure of polymers

Chemistry can be used to divide synthetic polymers into a series of groups that reflect different characteristics in behaviour. Although most readers of this book will be familiar with the principal types of plastics and rubbers, the nature of the chemistry is of such importance in determining the physical and technological properties that we revise it here as a lead-in to the structure–property relationships.

2.1   The method of synthesis

The chemical molecular structure of polymers depends on the method which is used to produce them.

A polymer chain is created either by an addition reaction to a double bond, via a condensation reaction, or by a metathetical (exchange) reaction. In all cases the final structure is a long chain, which has a high molar mass. In the case of synthetic polymers the chain will be formed by repeating the simple building block – a monomer – many times. The term polymer reflects the structure of the molecule: *poly* indicating that it involves many of the mono-*mer* units.

Ideally, studies and uses of polymer materials would involve samples in which all the polymer chains have the same chain length and hence the same molar mass. In practice, one rarely has a material with a unique chain length or molar mass and therefore it is important to understand how the distribution of molar mass can influence the physical properties.

So we start by looking at the various polymeric materials that can be obtained, but without going into the detailed mechanism of the synthetic processes involved.

2.2   Condensation polymers

The term "condensation polymers" covers all polymers created by the reaction of two organic functional groups, often with the elimination of a small molecule such as water. For instance, if we allow an alcohol to react with an acid, an ester is created (Figure 2.1).

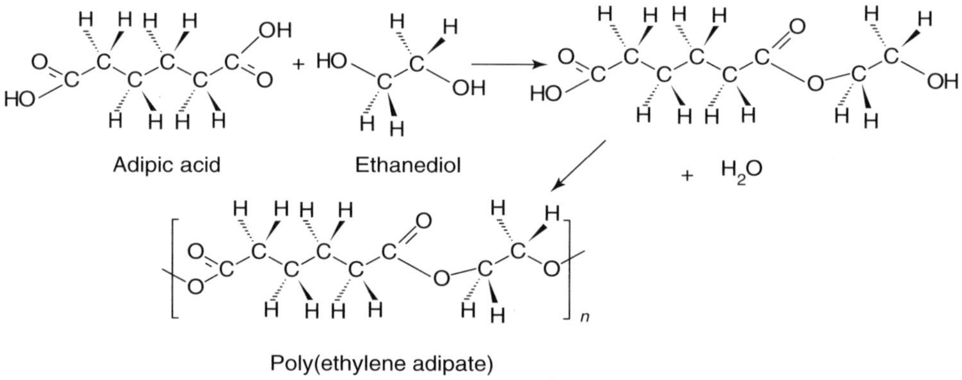

Figure 2.1   *Reaction of an acid with an alcohol to form an ester.*

In a similar way we can form ethers, amides, urethanes, ureas, sulfones and others. There are several important points to note in the formation of polymers by a condensation process:

(1) For a polymer to be formed, the reacting species must have at least two reactive functions per molecule.

(2) Reaction of 50% of the available functions will (on average) produce only a doubling of the molar mass of the material present. Each of these entities has the capability of undergoing further reaction but high molar mass polymers are only formed when virtually all the available functionalities are consumed by the reaction. To achieve a high molar mass polymer it is usual to find that the reaction had to go to 99.9% completion.

(3) Even when high molar mass material is produced, there will usually still remain reactive functions on the ends of the polymers. If monofunctional molecules are present, these can cap the polymers and stop further growth. Consequently, the presence of monofunctional impurities in the monomer mix is avoided if possible.

(4) Because of the step-wise addition reaction, the final material will possess chain lengths ranging from some with very great length to others which are barely longer than the dimers formed during the initial stages of the reaction. The precise form of the distribution of chain lengths in condensation polymers is of considerable importance in determining physical property variations.

Examples of the types of chemical functions which can be used to produce polymers are summarised in Table 2.1.

2.3   Addition polymers

A wide range of polymers are produced from addition reactions. The majority involve addition to a vinyl double bond and create a polymer which has a backbone that is

7

Table 2.1 *Examples of common chemical functions and the corresponding polymers formed*

Function	Structure	Function	Structure	Polymer	Structure
Acid	R'–C(=O)–O–H	Alcohol	HO–R"	Polyester	[R'–C(=O)–O–R"–O]$_n$
Acid	R'–C(=O)–O–H	Amide	H$_2$N–R"	Polyamide – Nylon	[R'–C(=O)–N(H)–R"]$_n$
Isocyanate	R'–NCO	Alcohol	HO–R"	Polyurethane	[R'–N(H)–C(=O)–O–R"]$_n$
Isocyanate	R'–NCO	Amine	H$_2$N–R"	Polyurea	[R'–N(H)–C(=O)–N(H)–R"]$_n$
Anhydride	(dianhydride structure)	Amine	H$_2$N–R"	Polyimide	[polyimide structure]$_n$

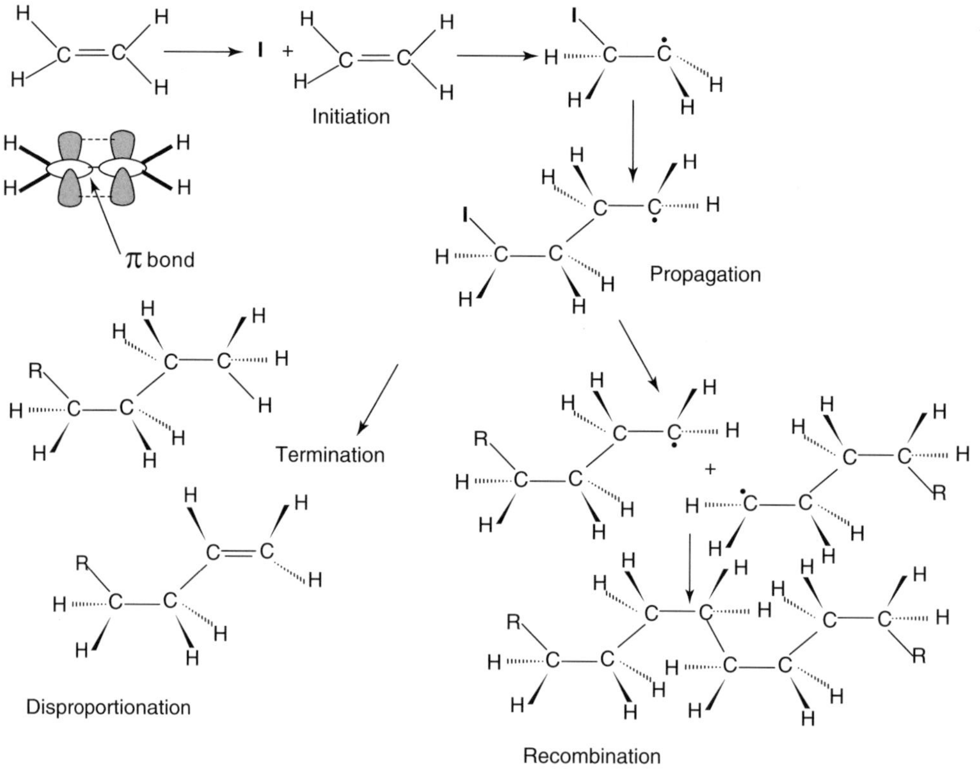

Figure 2.2 *Schematic diagram for the polymerisation of polyethylene.*

totally carbon–carbon bond linked. These polymers do not have the instability to exposure to strong acids or alkalis that typifies condensation polymers and so are preferable for many applications. Polymerisation is achieved by the presence in the monomer of a group that can either add or subtract electron density. Traditionally, polymerisation of vinyl monomers has been by free radical attack on the π electron cloud (Figure 2.2).

For example, ethylene (ethene) can react with a free radical to form a stable σ bond and a species which retains the free radical. This active site is capable of further reaction; propagation of the polymerisation will continue until the active radical site is destroyed. Table 2.2 contains a list of a number of monomers and the corresponding vinyl based polymers.

In the synthesis of such polymers the double bond can also be activated by use of an anionic or a cationic initiator. A characteristic of some of these latter materials is that they can have a very narrow distribution of chain lengths, in contrast to radical initiated polymers.

2.4   Stereochemistry in polymer molecules

In many fields of organic synthesis (such as drug development) it is necessary to control the stereochemistry (chirality) around the reaction centre. In the case of a

Table 2.2 *List of vinyl based polymers*

Monomer	Structure	Polymer common or IUPAC name	Structure
Ethylene – ethene		Polyethylene PE®	
Vinyl chloride – chloroethene		Poly(vinyl chloride) PVC®	
Propylene – propene		Polypropylene Propylene®	
Styrene – phenylethene		Polystyrene Styrene®	
Methyl acrylate		Poly(methyacrylate) Acrylic®	
Methyl methacrylate – methyl (2-methyl-propenoate)		Poly(methyl meth acrylate) Perspex®	

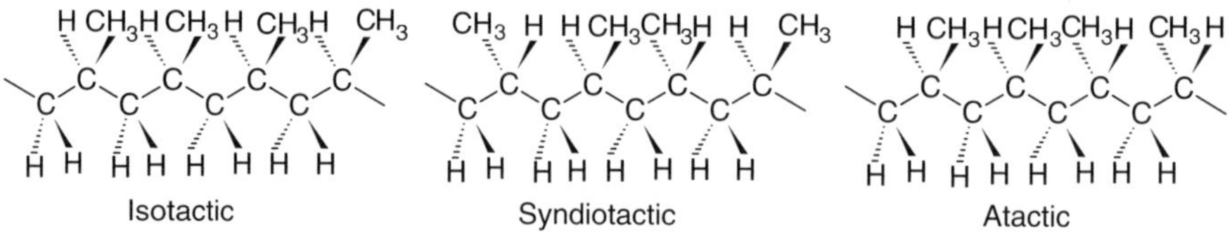

Figure 2.3 *Three tactic forms of polypropylene.*

radical initiated polymerisation this is difficult. However in the 1950s Zeigler and Natta discovered that it was possible to initiate the polymerisation of propylene using solid state catalysts. The catalysts created using aluminium trialkyl and titanium tetrachloride produced a polymer in which chiral control had been achieved at the reaction centre. So we now refer to the stereochemistry of the resulting polymer as its *tacticity*. If all the groups adding to the active centre give the same chirality, the polymer is termed *isotactic*. As an example we can consider polypropylene (Figure 2.3).

If the monomers add in an alternating manner, the polymer is termed *syndio-tactic*. Polymers without any control of the chiral addition, which is the case for normal free radical polymerisation, are termed *atactic*. The physical properties of polymers prepared with and without stereochemical control can be very differ-ent. For example, isotactic polypropylene is a hard crystalline material and can be used to make pipes used in domestic hot water systems, whereas the syndiotactic and atactic polymers are soft tacky materials, which are used as thickening agents for lubricating oils and as components in adhesives.

2.5 Geometric isomerism

In the case of important elastomers formed by polymerisation of conjugated dienes, addition across one double bond can, by electron migration, cause move-ment of the second double bond. This results in either 1,2 or 1,4 addition across the four-carbon monomer (Figure 2.4).

So polybutadiene can be obtained with different physical characteristics, depending on the conditions used in its synthesis. Simple addition to the first vinyl bond leads to the 1,2 addition product. This polymer is rigid at room tem-perature compared with the 1,4 addition product, which is rubbery. Furthermore 1,4 addition can create either *trans* or *cis* configurations, and these can have different degrees of local order.

2.6 Monomer functionality

A general rule which holds for many polymer systems is that if the monomer has two functional groups then a linear polymer results, but if the monomer

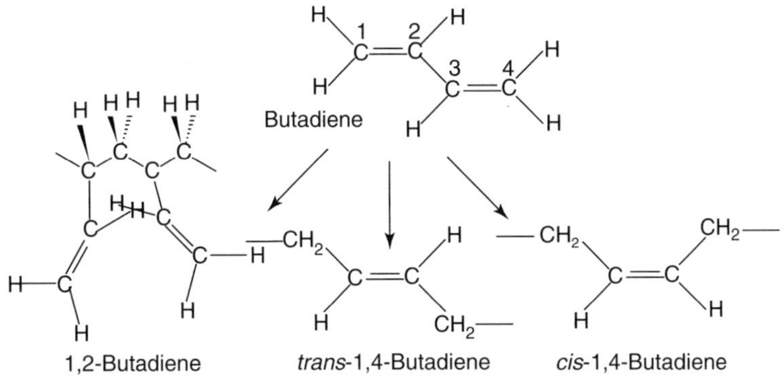

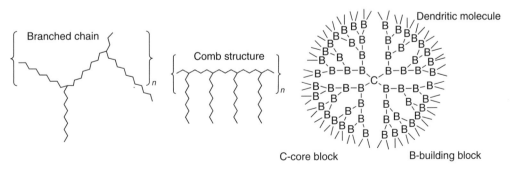

Butadiene

1,2-Butadiene   *trans*-1,4-Butadiene   *cis*-1,4-Butadiene

Figure 2.4 *Various configurations that can occur on polymerisation of butadiene.*

has a higher functionality then a branched chain polymer will be created. Conjugated dienes are an exception in that, although there are two double bonds per molecule, only one tends to be consumed in the initial polymerisation process. Other systems with functionalities greater than two will create either thermoset materials if the concentration of the multifunctional component is high, or branched chain materials if it is low. A thermoset is not converted to a free flowing melt on heating but retains the shape in which it was originally formed. Both addition and condensation polymerisations can yield such materials. Polymers with low concentrations of multifunctional monomers can still show thermoplastic properties, but the chains may be branched, comb or dendritic, depending on the level and nature of the branching (Figure 2.5).

The star-shaped polymers originate from a single polyfunctional point, producing a number of radiating arms. Because, in emanating from a single point, the chains there are very crowded together, normally it is not possible to synthesise such molecules with more than four arms. If the arms can then branch further, the molecules are called *dendritic* polymers. In the solid state these form

Branched chain

Comb structure

Dendritic molecule

C-core block   B-building block

Figure 2.5 *Schematics of various types of branched chain structures.*

what are called *dendrites*, where a dense clump of packed chains is surrounded by tassels of more loosely packed chain sections.

2.7 Copolymers and blends

The simplest way to produce a material in which the chemical composition of the material has been changed would be to blend two different types of polymer together. The polymer formed from a single monomer is termed a *homopolymer*. Whether or not blending results in a change or a superposition of the physical properties of the two component materials depends on whether or not they are thermodynamically compatible. If the polymers have similar structures then they may form a homogeneous mixture in the melt, and the physical properties which result are a blend of those of the component polymers. If, however, they do not mix then the resultant material is a mixture of separate phases, each of which will exhibit its own properties.

An alternative way of achieving physical property changes is to combine different monomers through synthesis, a process called *copolymerisation*. A range of species can be created using this approach, and each has its own particular characteristics. First we shall consider two monomers capable of being polymerised simultaneously.

(1) *Random copolymers*

   If monomers A and B are mutually soluble in one another, or in a common solvent, then they can be randomly incorporated into the backbone of the resulting polymer chain. Since there is essentially only statistical control of how the monomers are incorporated, the product contains a random distribution of monomers and can be depicted as follows:

   ABBAAABABABBBAAAABAABBABBABAABAAA

   The physical properties of the polymer reflect the averaged nature of the structure and so yield a material with properties intermediate between those of the two homopolymers.

(2) *Alternating copolymers*

   In certain cases, the monomers A and B, in the presence of a suitable catalyst, are incorporated into the polymer backbone in a regular alternating manner and the structure formed may be depicted as follows:

   ABABABABABABABABA

   The detailed structure of the elements of the polymer backbone is not that of either of the homopolymers. So the physical properties of this material are different from those of its constituents.

(3) *Block copolymers*

Using certain catalysts, it is possible to create a so-called living polymerisation process in which the active centre remains live after consuming all the monomer and can then add additional monomers in a sequential manner. These reactions can yield di- and tri-block copolymers:

AAAAAAAAAAAAAAAAAAAAAAAABBBBBBBBBBBBBBBBBBBB

Diblock copolymer

AAAAAAAAAAAAAAABBBBBBBBBBBBBBBBBBBAAAAAAAAAA

Triblock copolymer

Since most polymers are not miscible, the elements of the chains separate into two phases and the resultant material has some of the properties of A and some of B. The possible morphologies are:

(1) Spheres of A dispersed in a matrix of B.
(2) With extruded materials the morphology is anisotropic; there can be cylinders lying along the extrusion direction and having circular sections transverse to the extrusion direction.
(3) Lamellar structures can be obtained in some materials.

If A is a hard material and B is an elastomer then materials with highly anisotropic moduli may result from extrusion. These will have a high modulus in the extrusion direction due to the extended cylinders or sheets of the harder material, but a lower modulus perpendicular to the extrusion direction where the harder phases are separated by the elastomer. Since these materials have a phase structure on the nanometre scale, they can be considered as molecular nanocomposites.

So the nature of the morphology created depends on the thermodynamics of mixing of the two polymers and also on the chain lengths between the changes in chemical type.

2.8 Local organisation of polymer chains

In the simplest polymer, polyethylene, the chains can come close together because the backbone is not substituted. So the material can exhibit an ordered and therefore crystalline structure. This is illustrated in simplified form in Figure 2.6.

This imparts to the polymer a distinct crystal melting transition and in that respect it is similar to low molar mass organic crystals. Not all the sections of the chain are organised in this way. Some of the chains are disordered and the implications of this will be considered further in Chapters 6 and 9. These partially

Figure 2.6 *Schematic of the order observed in polyethylene. The zigzag represents the carbon–carbon backbone of the polyethylene chain.*

crystalline materials have domains that can possess different refractive indices, as well as having dimensions of the order of the wavelength of light. These domains scatter light to give the material an opaque appearance. On the other hand, if the polymer molecules contain bulky groups as side chains, packing is inhibited and the material will be unable to create an ordered crystalline phase. Such materials cool to form disordered glasses that are termed amorphous. Amorphous materials have a uniform refractive index throughout, so will not scatter light and are usually transparent plastics.

So we can subdivide polymer materials according to the categorisation shown in Figure 2.7.

Although in principle it is possible to create an ordered thermoset material, generally the polymerisation of a multifunctional monomer will create an amorphous, disordered structure.

The molar mass will change depending on the number of monomers which are incorporated into the polymer chain. A subscript n is usually used in formulae to indicate the number of monomer units in one polymer chain, e.g. as in Figure 2.1. This subscript is referred to as the degree of polymerisation. For low values of n, the materials are called *oligomers* or *telomers*. As the value of n increases, the physical properties change in a systematic manner. Thus molar mass is one of the parameters that commercial producers can vary in order to make

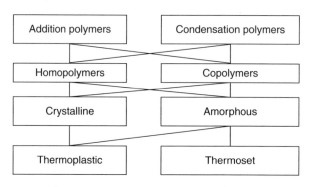

Figure 2.7 *Categorisation of plastics.*

the material more easy to process or in order to have enhanced physical properties. It is not unusual to find that a particular homopolymer may be available in
seven or eight different grades, each with slightly different characteristics reflecting the blends of chain lengths in the material.

2.9 Molar mass distribution

We have seen that, depending on the method used for synthesis, the chain length
may vary and hence the physical properties may also vary. If, as in the case of
condensation polymers, significant numbers of short chains are retained in the
final material then these materials do not exhibit sharp changes in their physical properties. In contrast, addition polymers can have less material of low chain
length and thereby exhibit narrower chain length distributions and sharper variations in physical properties. These differences are illustrated by the distributions
depicted in Figure 2.8.

In this figure we have introduced the fact that in a particular sample the measured molar mass is an average. Here we have used a *number average* defined as

$$M_n = \sum N_i M_i / \sum N_i$$

where N_i is the number of molecules of molecular mass M_i. An alternative
approach to the definition of the distribution is in terms of the weight (or
weighted) average molar mass distribution defined as

$$M_w = \sum W_i M_i / \sum W_i$$

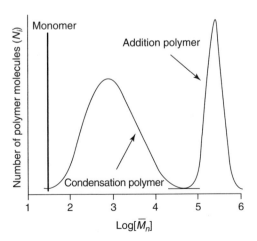

Figure 2.8 *Typical molecular mass distributions for vinyl and condensation polymers.*

where $W_i = M_i N_i$ and the molar mass distribution is defined as the ratio of M_w/M_n.

For a typical condensation polymerisation, the molar mass distribution function is generally in the range 3–20, but is sometimes even greater. On the other hand, in vinyl polymerisation the values typically will be in the range 1.05–3.0. The narrowest molar mass distributions are observed with anionic and certain cationic initiated polymerisations. Molar mass effects are observed with all polymer systems but they are more important in the physical properties of amorphous polymers than in their crystalline analogues.

2.10 Summary

In this chapter we have seen that:

- Different polymer types are obtained by different synthetic methods.
- These may be condensation polymers containing linking groups such as ethers, amides, urethanes, ureas, sulfones and others, or they may be addition polymers, usually with carbon–carbon linkages formed by addition across a double bond.
- Formation of the polymers may result in the creation of asymmetric centres, and the ways in which these chiral centres are arranged along the chain determines the overall optical isomerism of the chain, called the tacticity.
- Chains of different tacticity can have very different physical properties.
- Certain monomers, such as conjugated dienes, can give polymers of different geometric isomerism; these will also have different physical properties.
- If the monomer has functionality greater than two, the chain can become branched at that point; the resulting polymers range from slightly branched chains, to more and more heavily cross-linked chains, to an infinite network of cross-linked material.
- It is also possible to form star-shaped polymers with several unbranched arms radiating from a single point, or dendritic polymers with branched arms in the star.
- A wide range of different chemical structures, morphologies and physical properties can be achieved by polymerising more than one monomer type into the polymer chain; these copolymers may have random or ordered arrangements of the monomer types.
- Block copolymers usually form a two-phase morphology, with consequent unusual physical properties.
- Generally polymers are synthesised not with a single definite molar mass but with a distribution of different molar masses; again, the form of this distribution affects the final physical properties.

Further reading

Polymer synthesis

Cowie, J.M.G. and Arrighi, V. *Polymers: Chemistry and Physics of Modern Materials*, 3rd edn., CRC Press, Boca Raton, 2008.

Challa, G. *Polymer Chemistry, An Introduction*, Ellis Horwood, New York, 1998.

Stevens, M.P. *Polymer Chemistry, An Introduction*, 3rd edn., Oxford University Press, 1999.

Pethrick, R.A. *Polymer Science and Technology for Engineers and Scientists*, Whittles Publishing, Dunbeath, 2010.

Polymer characterisation

Pethrick, R.A. and Dawkins J.V. (Eds.) *Modern Techniques for Polymer Characterisation*, John Wiley & Sons, Chichester, 1999.

3

Nature of molecular motion in polymers

The dynamic behaviour of polymer molecules is the bridging element in understanding the relationship between the chemical structure of a polymer and its physical properties. Molecular movement usually involves some change in the conformation of parts of the polymer chain. Since many readers may not have had experience of conformational analysis, it is now appropriate to consider the various forms of motion which are possible for polymers by starting with their small molecule analogues.

3.1 Molecular motions in flexible polymers

The physical properties of flexible polymers are determined by two basic molecular motions. Just as with small molecules, the energy of a polymer molecule is defined by its shape and position in space. Since a polymer molecule is very large, we have to consider both the location of its centre of gravity, and its shape in terms of the distribution of the monomer units about the centre of gravity, called its *conformation*. The first motion we shall consider is this change in shape of the molecule, a conformational change brought about by internal rotation about the backbone bonds of the polymer chain. The second is translation of one molecule relative to another. These are illustrated in Figure 3.1.

Consideration of the shape change in polymer molecules requires us first to discuss the conformational changes in small sections of the chain, and then to consider the way in which the chain moves as a whole. Change of shape of the chain as a whole can result in translation, which is effectively diffusion.

3.2 Internal rotation in small molecules

To understand the complex conformational changes which occur in polymer molecules, we start by considering the processes in small molecules. Then we see how these are changed by progressively lengthening the chain.

We first examine the relationship between conformation and energy. Rotational isomerisation in simple molecules has been extensively studied, and values for the energy differences between various forms, as well as the activation energies opposing interconversion, have been measured. To understand these

19

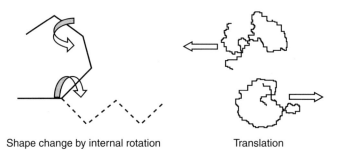

Shape change by internal rotation   Translation

Figure 3.1 *Simplistic picture of two important molecular motions.*

energies it is appropriate to start with the nature of the forces that control the interactions between molecules and their shape.

3.3   The interactions between molecules

We start with the relative strengths of non-bonding interactions in molecular systems.

In a molecule such as methane, CH_4, the hydrogen atoms are bonded to the carbon through sharing of electron density. The electron density is not equally distributed between the two atoms and there is a small charge imbalance, which results in a small bond dipole of 0.54 D. However, since the bonds in methane are tetrahedrally disposed to one another, there is no net dipole moment and the molecule as a whole is neutral. Looking at the molecule from afar, we see only a very small positive charge on each of the hydrogen atoms and a corresponding negative charge on the carbon. The small charge imbalance is able to attract other molecules and is the source of the attractive forces which ultimately allow methane to condense to a liquid and, at lower temperatures, to a solid. If two molecules are brought very close together, there is a large repulsive force. The balance between the attractive and repulsive forces gives rise to an energy minimum at a characteristic separation (Figure 3.2). This energy minimum will determine the equilibrium distance between the molecules in the solid phase and the temperature at which the solid melts.

While methane is a non-polar molecule, chloromethane, CH_3Cl, has a dipole moment of 2.1 D as a consequence of polarisation associated with the C–Cl bond. The larger negative charge on the chlorine atom will have a bigger attraction for the positive charged carbon, and results in a larger value of the interaction energy with neighbouring molecules. As a consequence, chloromethane has a higher melting point than methane. Attractive interactions between dipoles can operate over longer distances than interactions between non-polar molecules, and they have profiles similar to those in Figure 3.2 except that the potential energy minimum is deeper and the distances involved are greater. If a formal

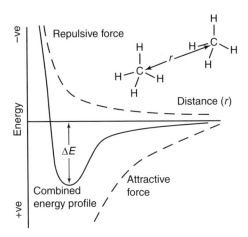

Figure 3.2 *Schematic representation of the energy distance profile for two molecules interacting; ΔE represents the stabilising interaction energy.*

charge is present then attractive interactions operate over even larger distances and the energy minimum is proportionately deeper.

Bringing a dipole or a charged species close to a non–polarised bond can induce a polarisation of the electron density, and a small dipole is induced. This small transient dipole can now interact with the permanent dipole and a larger energy minimum is observed than in the absence of this additional interaction.

Most of the contributions which influence the rotational energy profile are non-bonding; in other words there is no electron transfer involved in the interactions. There is one important exception, and that is hydrogen bonding. Hydrogen bonds are linear and involve a combination of a charged and a non–charged state (Figure 3.3).

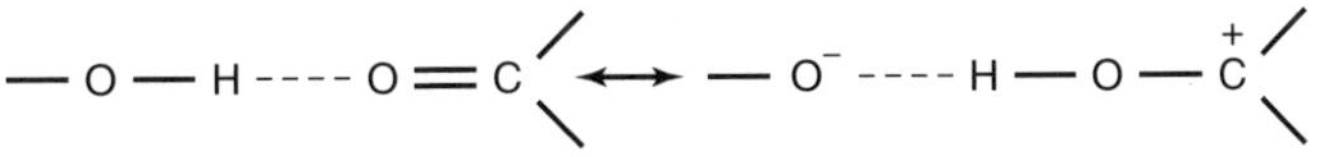

Figure 3.3 *The states involved in determining the hydrogen-bonding interaction.*

The hydrogen can rapidly exchange between the two core molecules and hence forms a bond which is stronger than the Van der Waals, dipolar or electrostatic interactions, although it is weaker than formal σ or π covalent bonds.

3.4 Internal rotation in small and larger molecules

The forces which determine the shape of the rotational isomeric potential are predominantly non-bonding interactions, similar to those described above, but now between atoms attached to neighbouring bonds.

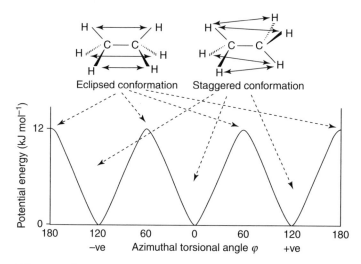

Figure 3.4 *The saw-horse representation, potential energy curve and indication of principal non-bonding interactions for ethane.*

These interactions are very short range and fall off as a high power of the distance. The interactions for the eclipsed and staggered structures of ethane are shown in Figure 3.4.

The arrows indicate the non-bonding interactions presented in a "saw-horse" representation. Only the strongest interactions have been shown; however, there are longer, and hence weaker, interactions between all the other hydrogen atoms, as well as the strong bonding interactions which retain the tetrahedral symmetry of the methyl groups.

The rotational isomeric structures for ethane have two extreme cases; the lower energy staggered and the higher energy eclipsed forms. In the eclipsed form the distance between the hydrogen atoms on adjacent carbon atoms is significantly shorter than in the staggered case, and this is reflected in a higher potential energy.

Throughout our discussion of polymer molecular motion we shall repeatedly refer to the potential energy rotational isomeric diagram. This is shown in a simple form in Figure 3.4. As we move to a more complex molecule such as butane, the diagram represents the way in which the energy of the molecule changes as it is converted from the *trans* (or transoid) to the *gauche* (or cisoid) form. The majority of the molecules will be in either the *gauche* or the *trans* state. A negligible number will be found in the eclipsed states undergoing transition between the two lower energy forms. Therefore the total conformational energy is that of molecules distributed between these *gauche* and *trans* states. If the temperature is raised then according to the Boltzmann distribution some of the molecules in the lower energy *trans* state will be transferred into the higher

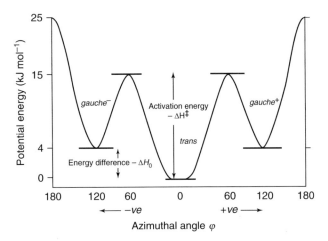

Figure 3.5 *Energy diagram for rotational isomerism of butane.*

energy *gauche* state. The rate at which this occurs will depend on the height of the barrier between states as defined by the potential energy of the eclipsed state. This occurs at an azimuthal angle of $\pm 60°$ and is illustrated in Figure 3.5.

The next member of the isomeric series of hydrocarbons is pentane. If rotation of the terminal methyl groups is ignored, there are two carbon–carbon bonds to be considered (Figure 3.5). Whereas in the case of ethane and butane it is possible to map the change in energy in terms of a two-dimensional curve as shown in Figure 3.4, for pentane and longer chain molecules it is necessary to consider a potential energy surface. The stable conformations which are possible are shown in Figure 3.6.

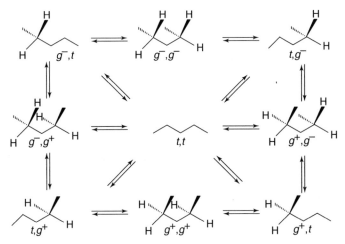

Figure 3.6 *Conformations of pentane.*

The lowest energy conformation is one in which all the bonds are in a *trans* conformation (*tt*). The energy profile can be represented as a three-dimensional contour plot with φ_1 and φ_4, the torsional angles of the terminal methyl groups, equal to zero. We assume that the rotational isomerism involves the bonds between carbon atoms (2) and (3) with torsional angle φ_2, and carbon atoms (3) and (4) with torsional angle φ_3. In this case, when two bonds are simultaneously involved in conformational changes, the potential energy can be expressed as a function of multiple torsional angles in multidimensional space, often referred to as "*conformational space*". Figure 3.7 represents the contours of the conformational space for pentane.

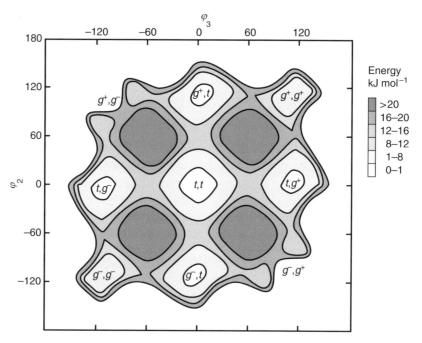

Figure 3.7 *Conformational space for pentane.*

Table 3.1 *Energies of the conformations of pentane*

Conformation	Energy (kJ mol⁻¹)	Conformation	Energy (kJ mol⁻¹)
t,t	0	*t,g⁺*	2.1
t,g⁻	2.1	*g⁺,t*	2.1
g⁻,t	2.1	*g⁺,g⁻*	7.5
g⁻,g⁺	7.5	*g⁻,g⁻*	9.6
g⁺,g⁺	9.6		

If we take a line between the two conformations, the potential energy profile is essentially similar to that shown in Figure 3.5 for butane. The energies of the conformations are shown in Table 3.1.

The t,t state is the lowest energy state and is taken as the reference for all other states. The conformations created by rotation about either φ_2 or φ_3 are $t,g^- = g^-,t = g^+,t = t,g^+$. The intramolecular forces defining these conformations are the same, making these conformations energetically degenerate. Rotation about both φ_2 and φ_3 produces the conformations $g^+,g^- = g^-,g^+$, which are also energetically degenerate, but involve stronger interactions than those in the creation of one *gauche* conformation. The sterically most crowded conformations are the g^+,g^+ and g^-,g^- structures. The close proximity of the methyl groups must cause their rotation to be coupled to the backbone rotation about torsional angles φ_2 and φ_3. There are two energy states slightly displaced from the expected 120° location (Figure 3.7), which implies that the equilibrium conformations are imposing a slight twist on the carbon backbone. As we shall see later, such small effects can have a profound effect on structures adopted by polymers in the solid state.

Extending the chain length makes it difficult to visualise the conformational potential energy space as the dimensionality should increase with the number of bonds in the molecule. However, if it is assumed that the important intramolecular

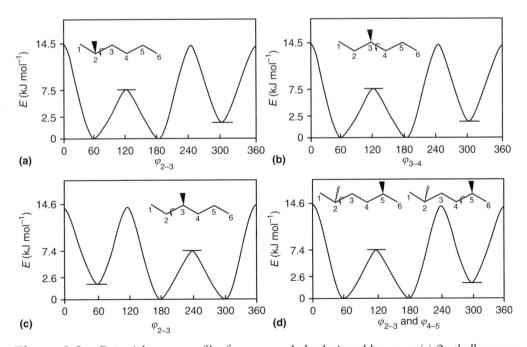

Figure 3.8 *Potential energy profiles for some methyl substituted hexanes: (a) 2-ethylhexane; (b) and (c) 3-methylhexane; and (d) 2,5-dimethylhexane.*

interactions act over only very short distances, then it is possible to simplify the considerations to only the sterically hindered portions of the molecule.

Increasing the chain length of the alkanes, conformational analysis of a series of methyl substituted hexanes indicates that the important potential energy profiles can be identified with those bonds which contain the methyl substituent. This has important consequences for polymers. The theoretical rotational isomeric potential profile for 2-methylhexane is shown in Figure 3.8.

Considering the conformations associated with the carbon(2)–carbon(3) bond, there are two lower energy forms and a single higher energy form. This is a different distribution from that found in butane, where the lowest energy state is singly degenerate and the upper is state is doubly degenerate.

Changing the position of the methyl substituent changes the degeneracy of the upper and lower states, but interestingly does not have a major effect on the energy difference of the barrier to internal rotation. Because of the energy difference between the conformers it is possible to measure the rate of exchange, which of course is related to the barrier to internal rotation (Table 3.2).

In 2,5-dimethylhexane the values of the barrier and energy difference are very close to those of 2-methylhexane, indicating that the process being observed can be considered as independent rotation of the C(2)–C(3) and C(4)–C(5) bonds. Hexane can exist in the extended all *trans* conformation or can contain a number of *gauche* conformations. Two of the many conformations it can adopt are presented in Figure 3.9.

The all *trans* conformation is the lowest energy state, but the conformations that contain progressively more *gauche* forms adopt a structure that closely resembles that of a cyclohexane ring. This type of folded structure plays a significant role in the crystallisation of hydrocarbons.

Table 3.2 *Comparison of experiment and theory for the rotational dynamics of methylhexanes*

Molecule	Dihedral angle (φ)	$\Delta H^{\ddagger}_{exp}$ (kJ mol^{-1})	$\Delta H^{\ddagger}_{theory}$ (kJ mol^{-1})	ΔH^{0}_{exp} (kJ mol^{-1})	ΔH^{0}_{theory} (kJ mol^{-1})
2-Methylhexane	C(2)–C(3) 60, 180→300	11.3 ± 2	12	3.6 ± 1	2.5
3-Methylhexane	C(2)–C(3) 60, 180→300 C(3)–C(4) 60, 180 → 300	13.2 ± 2	12	2.9 ± 1	2.6, 2.5
2,5-Dimethylhexane	C(2)–C(3) and C(4)–C(5) 60, 180 → 300	14 ± 2	12	1.2 ± 1.3	2.5

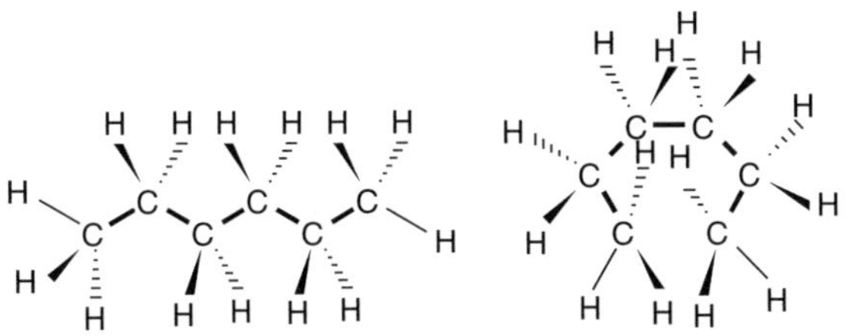

Figure 3.9 *Two of the possible conformations for hexane.*

The challenge for the polymer scientist is to understand how these very simple rotational isomeric processes in small molecules become incorporated into the dynamics of a long polymer chain.

3.5 Rotational isomeric state model of polymers

Paul J. Flory addressed the problem of predicting the shape of a polymer molecule and so provided an understanding of the way in which chemical structure influences conformational changes in polymers. His approach was essentially an extension of that presented above for pentane and the substituted hexanes. He considered that the range of the important interactions was very short. This approximation is surprisingly accurate and has allowed theoretical prediction of the equilibrium structures of many polymers, at least when the molecules are present as separate entities in dilute solution.

We start with the simplest polymer, polyethylene. Figure 3.10 shows a section of the chain in an all *trans* conformation.

The conformation of the polymer can be described in terms of the position vectors of the backbone. For each bond there will be a value of φ which indicates whether that bond is in a *gauche* or *trans* conformation. If we have $N + 1$ monomers then we have $N + 1$ position vectors: $R_0, R_1, R_2, R_3, R_4, \ldots R_N$. We then have $N - 1$ bond vectors

$$r_0 = R_1 - R_0, \quad r_1 = R_2 - R_1, \ldots r_{N-1} = R_N - R_{N-1}$$

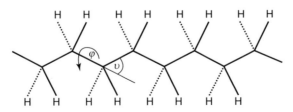

Figure 3.10 *Saw-horse representation of polyethylene, omitting backbone carbon atoms.*

and $N - 2$ dihedral angles: $\varphi_0, \varphi_1, \varphi_2, \ldots \varphi_{N-2}$. The dihedral angle φ_i is the angle between the plane of the vectors r_{i-1} and r_i, and the plane of the vectors r_i and r_{i+1}. In order to completely specify the conformation we should also give the $N - 1$ angles and the N bond lengths. This description leaves six variables to fix the centre of mass and the orientation of the molecule.

It is usually assumed that the potential energy surface for polyethylene is essentially the same as that for butane (Figure 3.5). Since the range of the forces is relatively short, the potential energy surface for the polymer does relate very closely to that of the simple butane analogue. This assumption then incorporates the "pentane effect", in which the g^+, g^+ and g^-, g^- conformations would bring the chains almost into the same region of space. The mathematical analysis of the distribution of conformations includes this, but does not allow for the possibility that some other chain sections attempt to occupy the same region. This is the *"excluded volume"* effect.

Changing the intramolecular interactions produces additional features in the potential energy profile. These can have a profound effect on the solid state crystal structure. This can be seen in the case of the replacement of a carbon by an oxygen atom, as in poly(methylene oxide), $-(CH_2O)_n-$, and the related system poly(ethylene oxide), $-(CH_2CH_2O)_n-$. Halogen atoms, and in particular fluorine, introduce the possibility of a longer range dipolar component in the intramolecular interaction energy. As a consequence of the longer range of the forces, the potential energy profile becomes sensitive to the conformation of bonds which are further from the bond about which rotation is taking place. Many polymers contain aromatic ring structures, carbonyl groups and hydroxyl groups and these introduce the possibility of further induced dipole interactions and hydrogen bonding.

Polymers are large molecules and the degree of polymerisation, i.e. the number of monomers incorporated in the chain, will typically range from 100 to greater than 1000. The number of possible conformations for such a molecule would be around 2.0×10^{26}, a very large number indeed. Yet, despite these complications, we can still use this analysis to obtain reasonable estimates of the potential energy surfaces and thence predictions of the conformational dynamics of polymer chains.

If it is assumed that the chain is flexible, that is that the barrier between conformational states is sufficiently low that rapid interchange can occur, then the time averaged shape of the polymer will be described by the distribution between the available conformations. This allows a statistical mechanical analysis, called the rotational isomeric model. This indicates that the population of the *gauche* states causes the chain to adopt a random quasi-spherical shape and not the often pictured zigzag extended all *trans* form. The analysis allows calculation of two fundamental parameters: the radius of gyration, which is the size of the polymer molecule if it were to undergo free internal rotation, and the related end-to-end distance. These quantities define the shape of the isolated polymer when it is dispersed in dilute solution and indicate that it is rather like a ball of string. One very important observation, which arises from this analysis and is confirmed by experiment, is that as the temperature is increased, so the size of the polymer molecule shrinks! Normally we consider that when we heat materials they will expand, and this is what we observe for bulk samples of most plastics. This is because of expansion of the free volume between the molecules. However, at a molecular level polymer chains will coil up as the chains attempt to incorporate more of the higher energy kinked *gauche* conformations.

The success of the rotational isomeric model lies in predicting the time averaged shape and size of polymer molecules in dilute solution. However, we have to look further to find the way in which the molecule moves between the various states that are averaged in the model.

3.6   Dynamics of conformational change

In principle it is possible to simulate the motion behaviour of a polymer using the calculations of theoretical molecular dynamics. These calculations involve randomly changing a *trans* conformation to *gauche*, or vice versa. The rate at which the change takes place is dictated by the activation energy (Figure 3.4). Such calculations immediately run into problems; for a sensible polymer chain we need to include 100 monomers and, as indicated previously, this requires consideration of 10^{26} possible conformations. Reducing the chain length to 30 still leaves the possible number of conformations at 3.5×10^7, which is a large matrix to handle. Attempts at theoretical calculations for polymers are therefore usually restricted to relatively short chains of 10–20 or so monomer units.

Since we are concerned with the dynamics, we have to ask how long it takes for a monomer unit to hop between the various states. The time can be of the order of 10^{-9} to 10^{-6} s. So the step interval in the computation is often taken as 10^{-10} s to allow the changes to be accurately mapped. This is extremely laborious and so an alternative approach is required. We achieve this with a molecular kinetic model.

If we consider the case of butane (Figure 3.5), the hopping can be considered as a simple kinetic process, and the rate constant then reflects the time for the interconversion between states. Note that two relevant rates of interconversion have to be considered: the rate at which the *trans* conformation is converted into the *gauche* conformation, over the energy barrier denoted as $\Delta H^{\ddagger}_{t,g}$, and the reverse rate for the *gauche* to *trans* conversion, for which we designate the barrier $\Delta H^{\ddagger}_{g,t}$. The two quantities are related by the energy difference between the *trans* and *gauche* states, ΔH_0

$$\Delta H^{\ddagger}_{t,g} = \Delta H^{\ddagger}_{g,t} + \Delta H_0$$

Kinetic studies of the interchange process allow rate and energy values for specific changes to be measured. In the usual case, where there is more than one higher energy state, an average value is observed. Of course, if the technique can identify a specific conformation, then an exact value can be measured. However, in the case of a polymer molecule, average values are observed.

If we consider an element of the polymer chain undergoing a conformational change from a *trans* to a *gauche* state, the moving element of the chain has to pass through a significant distance in space (Figure 3.11). For simplicity, the hydrogen atoms have been omitted from this representation of the polyethylene chain. This process looks much hindered except at the very ends of the chain. This immediately points to the possibility of chain end effects in the motion dependent properties of the polymer.

The movements which are executed by a polymer chain will resemble those depicted as a crankshaft-like motion in Figure 3.11.

This does not require the impossible movement of a large element of the chain through a large distance, but it does involve the concerted motion of a number of bonds. In the example drawn above, the conformational change occurs with the cooperative motion of eight bonds. This is likely to be slower than isomerisation of the end group, which only involves rotation about a single bond. The motion also requires a small lateral movement. Since the chain is not initially in an all *trans* conformation, the process may involve a loop moving from one point in space to another. So the question is how many bonds, on average, are involved in the moving loop and how much energy does this require? Careful studies of the activation energy for rotational isomerism of polymers in dilute solution indicate that it is much larger than would be expected from the simple potential energy profile indicated above. So the fundamental conclusion is that the cooperative motion of a number of bonds requires more energy than rotation around a single bond.

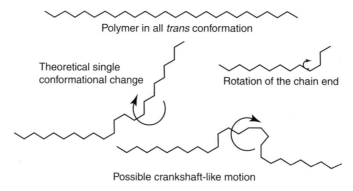

Polymer in all *trans* conformation

Theoretical single conformational change

Rotation of the chain end

Possible crankshaft-like motion

Figure 3.11 *Possible rotational isomeric motion of a polymer chain.*

Alternatively, if we were to attempt to achieve rotation about one single backbone bond, the motion would require a large element of the polymer chain to move through space. This is very unlikely to occur. So we next look at the coupled nature of the motions in polymers.

3.7 The coupled nature of internal rotations in polymers

Ignorance of the way in which motions are coupled presents a significant difficulty in visualising the molecular motions in polymer systems. However, comparisons with the small molecule analogue of polystyrene, 2,4,6,8-tetraphenylnonane, showed that in polystyrene the preferred conformation of the syndiotactic chains is with diads which are *trans-trans* (*tt*) and *gauche-gauche* (*gg*). The *gg* sequences create an approximately 90° change in direction of the backbone (Figure 3.12).

Since small molecule studies show that steric interactions prevent the occurrence of two adjacent *gg* states, the chain can be represented as a zigzag projected onto the planes of a square and would have the conformational structure in which state (1) changes into state (2):

$$t.gg.tt.gg.tt.tt \Longleftrightarrow t.tt.tt.gg.tt.gg$$

The process of translation of state (1) to state (2) has effectively moved two *gg* arrangements along the chain. For this cooperative motion to occur, there is also a requirement that sufficient intermolecular space should be available. Note that the end-to-end length of the chain has not been changed and the number of conformations with *gg* and *tt* sections is likewise unaltered.

An alternative process, which produces a change of four monomer units, is depicted in Figure 3.13.

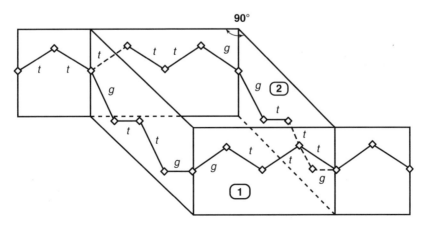

Figure 3.12   *Possible motion of a syndiotactic polystyrene chain.*

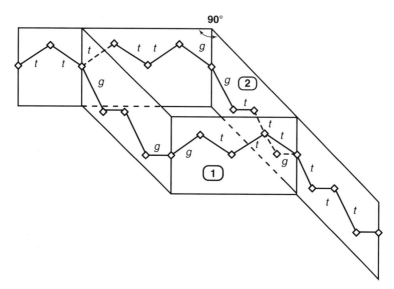

Figure 3.13 *Alternative possible motion of a syndiotactic polystyrene chain.*

Using the same conformational analysis as for the previous sequence, we can describe this process as:

$$t.tt.gg.tt.gg.tt \Longleftrightarrow t.tt.tt.gg.tt.tt$$

This causes not only a change in the direction of the chain, but also a change in the number of *tt* and *gg* conformations. This corresponds to a straightening of the chain and would therefore be observed when a mechanical stress is applied. In a real chain, a mixture of both the above processes is possible, as well as simple rotational isomerism at the chain ends. As the chain length increases, so the motion of the chains will become more constrained and the relative importance of these different processes will change.

3.8 The size of the moving element

The energy difference associated with the above process in polystyrene is about double the value of the same movement in 2,4–diphenylpentane, confirming that in the polymer this type of motion involves a number of monomer units.

The local potential surface involving five monomer units (ten backbone bonds) has ten conformational angles and ten valence angles along the main chain. Small deviations in these angles do not involve the same energy change as the complete rotation, and so the total energy involved is not ten times that for a single bond rotation in a small molecule, although it is significantly higher. Illustrating this, the activation energy for short chain polystyrene is approximately 17 kJ mol⁻¹, whereas for a high molecular weight polymer, in which a larger number of elements become involved, it is 35 kJ mol⁻¹.

A series of copolymers of an alkane chain with a styrene dimer or an α-methylstyrene dimer illustrate the way in which the conformational changes become easier as the distance between the substituent phenyl rings is increased. When the alkane chain length $n = 0$, the materials are polystyrene and poly(α-methylstyrene). Increasing the size of the alkane block in the copolymer reduces the interaction between the phenyl groups in the adjacent dimers and this is reflected in a lowering of the activation energy for the conformational change. When the alkane separation of the phenyl groups is small, movements in the two blocks are cooperative. However, when the alkane block becomes greater than about six carbon atoms, the activation energies for the conformational changes of the aromatic block and of the alkane become increasingly independent as their motions become decoupled.

To summarise: in polymer systems conformational change does not involve independent motion of groups around a single bond, but will involve the concerted motion of between six and ten bonds. The only situation where rotation about a single bond will occur is if the moving group is not attached at both sides, as in a side chain, or at a polymer chain end.

3.9 Libration

Very often the available energy is not adequate for full 360° internal rotation. Under these circumstances two processes requiring less energy may be able to take place. The first of these is partial rotation to some state higher in energy than the lowest energy state, but still lower in energy than the energy barrier of the eclipsed state. One such state might be the *gauche* conformer. Such a partial rotation, to and fro about the lowest energy point, is called *libration*. We call the activation energy for this process $\Delta E_\beta^\ddagger$, for reasons that will become clear when we examine the mechanical properties of glasses (Figure 3.14).

A second molecular motion that might be able to take place at low temperatures is rotation of, or inside, the substituent side group on the chain. Very often, too, the libration and side group rotation are coupled, so that both occur at the same temperatures and require the same times.

3.10 Normal modes of motion

So far we have analysed the shape change of polymer molecules in terms of conformational changes in small rotating sections of the chain. However, it is also possible to analyse the deformations of the whole chain. When the temperature is

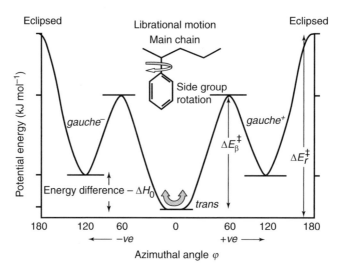

Figure 3.14 *Libration and side group rotation.*

sufficiently high for extensive conformational change to be possible, the molecule becomes what is called a *freely rotating chain* and then its local chemical structure becomes unimportant in determining the gross chain dynamics. This gives us a way of determining the movements of a chain in phenomena like flow and creep without including the details of chemical structure at the monomer unit level.

The first such treatment considers the coiled macromolecule as a soft ball. This becomes particularly useful in the discussion of the viscosity of polymer melts and solutions. To do this, we follow the work of Rouse and of Zimm, who returned to some observations by Isaac Newton.

Newton was studying the behaviour of a soft deformable ball in a shear field. He found that the ball was always deformed along the shear direction, but at the same time it was rotated, just like rolling a ball along a table (Figure 3.15).

If we now sit on the axes of the rotating deformed ball, and consider the original horizontal axis, it oscillates from long to short as the ball is rolled along. Correspondingly, the original vertical axis alternates between short and long. The net result is that as we go round with the ball it exhibits a "breathing"

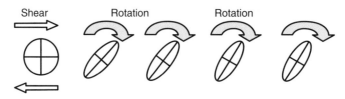

Figure 3.15 *Deformation of a soft ball under shear.*

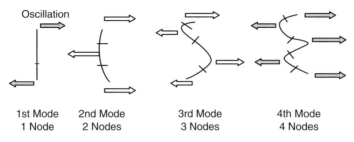

Figure 3.16 *Normal modes of motion.*

motion, going from "tall and thin" to "short and fat". Newton pointed out that this "breathing motion" is the result of a number of a fundamental and overtone movements. We call these *"normal modes"*, and they are characterised by the fact that all the sections of the ball move in phase. Graduates of chemistry are familiar with this definition because it applies to molecular vibrations as evidenced in infrared spectra.

So, just like the familiar vibrations of a string, the modes are characterised by how much of the ball moves in each direction. If we represent the ball by a line, and a node of no motion by a small cross-line, the various modes are illustrated in Figure 3.16.

Rouse and Zimm pointed out that we can imagine a polymer molecule as a Newtonian soft ball, at least when considered in isolation. Then the movements that take place, and absorb work energy under shear, are just these normal modes. The spectrum runs up to the point where the section between nodes is just a single equivalent freely rotating unit. Now the first mode has to move half of the chain in each direction. This is the heaviest task and involves movement through the biggest distance, so it requires the most energy and the longest time. The successive modes involve progressively shorter pieces of the chain, moving over progressively smaller distances. As a result they require less energy and time.

So long as there are enough freely rotating units in the chain to permit these modes of motion, the dynamic properties depend on the molar mass, but not the chemical nature, of the chain. We shall be treating this in more detail when we come to look at the property of flow viscosity.

3.11   Reptation in a tube

While the flow and creep of a polymer are necessarily associated with translational motion, the internal rotational shape change is taking place at the same time. So the total movement is a combination of the two. As before, when polymers approach the "ideal flexible" model with copious internal rotation, then their behaviour ceases to be sensitive to local conformational constraints and can

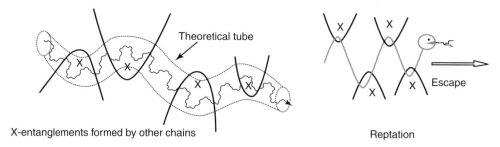

Figure 3.17 *Snake-like movement in a tube through entanglements.*

be modelled without reference to the chemical nature of the chain units. So the second "whole molecule" treatment considers the reference molecule by analogy with a simple worm-like or tube-like picture.

Our understanding of this double molecular motion in creep and flow follows from the work of Edwards and of de Gennes. In a bulk polymer material, the chains intertwine with each other in a series of *entanglements*. The translation of a reference chain is pictured as movement of a flexible molecule held within a tube, the geometry of which is constrained by entanglements with other chains or chain segments, as illustrated in Figure 3.17. This movement has been likened to the motion of a snake through the grass, moving forward while at the same time changing its shape around the grass stems. So, using the same classical word root as we find in "reptile", the motion is called *reptation*, while the overall picture is called a *tubular model*. Treatments of this model are complex because the "tube" constraining the "snake" or "worm" is not fixed in time and space, but is itself moving.

Melt flow and creep can occur when reptation has the energy and time to take place. But just how much time is needed for this composite motion? Since it occurs at the highest temperature of all glass–rubber–melt states, obviously it requires the most energy. In other words, the energy required to escape from an entanglement is greater than that required for simple conformational change. Then, having the highest energy barrier to overcome, and since the rate of a molecular process decreases with the height of the activation energy barrier, reptation is the slowest of all the movements we have considered so far.

From the fact that the high viscosity of a molten polymer varies as almost the power to 3.4 of the molar mass, reptation must have a similar high dependence on molar mass. But why is the process also so very dependent on the time available for it?

Reptation consists of two coupled processes (shape change and translation), both of which can be considered as diffusion random walks. Now the statistics of a single random walk tell us that the distance moved is proportional to the square root of the number of steps, and so to time

$$d=(2Dt)^{1/2}$$

where d is the distance moved and D is the diffusion coefficient for the random walk. For two processes to happen at the same time, the probability of both occurring together is the probability of the first multiplied by the probability of the second. Going from probability to distance moved in a random walk we get the square root of the square root;

$$d_{1,2}=(4D_{1,2}t)^{1/4}$$

So the distance diffused is proportional to the fourth root of time, or the time required is proportional to the fourth power of the distance to be moved. Now the time required is proportional to the viscosity, and the distance to escape is proportional to the chain length, or the molar mass. So the viscosity (and time for flow) of entangled chains should be proportional to the fourth power of the molar mass.

In fact, once part of the chain has escaped from its original entanglements, the rest of the chain may move more easily, and so the time and energy required are not quite as high as this simple treatment would show. The first simple treatment assumed that the entanglements, and so the tube, stay fixed in space, but this cannot be so as they are also moving chains. To counter this, early theories assumed that the motion of the polymer chain creates as many entanglements as those from which it has escaped. So the equilibrium number of entanglements and their effect is constant with time. However, experiments showed that this is not the case. Although more recent refinements of the theory allow for the entanglements to move, there are still problems with relating the viscosity to the 3.4 power of molar mass. Nevertheless the main precepts of reptation are still retained.

In conclusion, the important difference between the flow behaviour of short and long chains is the influence of chain entanglement, and this factor influences many of the physical properties, both in the melt and in the solid state.

Further reading

Theory
Doi M. and Edwards S.F. *The Theory of Polymer Dynamics*, Oxford Science Publications, Oxford, 1986.
Rubinstein M. and Colby R.H. *Polymer Physics*, Oxford University Press, Oxford, 2003.

Statistical mechanical–conformational approach
Mattice W.L. and Suter U.W. *Conformational Theory of Large Molecules*, John Wiley & Sons, New York, 1994.
Tonelli A.E. and Srivivasarao M. *Polymers from the Inside Out*, Wiley Interscience, New York, 2001.

4

The glass to rubber transition

4.1 The physical properties of interest

The great commercial impact that polymer materials have made is a consequence of their unique physical properties. Polymers are able to exhibit a combination of strength and flexibility which cannot be obtained with metals or ceramics. We encounter plastics either as hard crystalline or glassy solids, or as elastomers with rubbery mechanical characteristics.

Probably the most obvious physical property of a polymer is its hardness or softness. Indeed, the words "plastic" and "rubber" convey a picture of a deformable material. There are five important mechanical properties which form the core of our discussion. These are:

- hardness and softness and the ability to undergo reversible deformation;
- elastic recovery of original shape;
- vibration damping and energy loss;
- flow and "creep" – irreversible deformation;
- fracture – brittle and ductile breaks.

"Hardness" indicates how much force is required to bring about a rather small deformation, while "softness" indicates that a larger deformation is obtainable with a somewhat smaller force. However, in scientific studies it is necessary to be more precise about these quantities.

Starting with hardness, we define a scientific measure of this as the *modulus*. In essence, this is the force required to bring about unit deformation. The force is called *stress* and the deformation is called *strain*. Then modulus is the stress to bring about unit strain. There are different moduli depending on the geometric characteristics of the applied stress. The two most important of these are:

- linear elastic or Young's modulus;
- shear modulus.

The softness is quantified by the term *compliance*. This is the strain brought about by unit stress. It is, therefore, the mathematical inverse of the modulus.

Interestingly, when we come to the acoustic properties of polymer materials, we are concerned with almost the same properties as in mechanical studies. This is because the sound wave is just a high frequency pressure oscillation that is similar to a mechanical perturbation. Of course the important technological property is the *sound absorption*, and in Chapter 11 we shall discuss this using the same frequency dependent considerations as we do with energy damping or loss presented in Chapter 10.

The second most important use of polymeric materials arises from their excellence as electrical insulators. However, there are also low tonnage uses of plastic materials as electrical semiconductors and as organic electronic components in display technologies. So in this context, in Chapters 12 and 14 we shall examine:

- insulation and dielectric properties;
- capacitance and polarisation phenomena;
- conduction and semiconductor behaviour.

In fact, the electrical properties have many time and temperature dependent characteristics in common with the mechanical properties. The significant measure of the charging and polarisation (dielectric) behaviour of a polymer insulator is its *permittivity*. This can be thought of as a parallel to mechanical compliance where the stress is replaced by the electric field or voltage, and the strain by the charge movement or polarisation. Then the equivalent to mechanical loss is *electrical conductance loss* or *dissipation*.

In Chapter 13 we shall introduce a short section on some photo properties of polymers that are governed by phenomena similar to the behaviour presented above. The photo characteristics of polymers are becoming very important as they are the basis of the growing technology of organic light emitting diodes (OLEDs), which may very well displace liquid crystal displays in computers in the next few years. Here we concentrate on:

- light absorption and re-emission – fluorescence and phosphorescence;
- energy migration and transfer;
- energy trapping and stimulated emission.

In so doing we use the normal spectroscopic functions of *absorption coefficient* and *emission intensity*. These allow us to monitor the movement of energy as a quantum called an *exciton*. This energy can be transferred to another molecule behaving as a quenching agent, or it can be trapped by dimeric excited state complexes formed by the polymer and called *excimers* or *exciplexes*.

So in this chapter we shall start by examining the mechanical properties and see how they depend on temperature, time and morphology.

4.2 Mechanical properties

Four important mechanical properties – modulus and compliance, elastic recovery, vibration damping and energy loss, flow and creep – will be dealt with in this section.

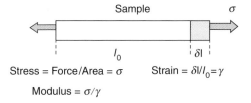

Figure 4.1 *Linear stress, strain and Young's modulus.*

The modulus can be defined either in terms of linear stress or shear stress. The linear, or Young's, modulus is shown in Figure 4.1.

When the stress is calculated as the applied force or load divided by the original cross-section area of the sample, it is called the *nominal stress*. This is what is measured in much tensile test equipment. However, when the sample is stretched, in order for the volume to remain constant, the cross-section area decreases. Then the applied force or load divided by the actual changing cross section area is called the *true stress*. In what follows, we shall use only Young's modulus when describing the mechanical properties of solid polymers. However, the essential property of a liquid is that it flows under the influence of a shearing stress. Consequently, when we come to look at molten or liquid polymers, we shall deal with shear moduli (Figure 4.2) and the related property of viscosity.

4.3   Four temperature regions of mechanical behaviour

A convenient basis for all that follows is obtained by looking at the modulus of a polymer material as a function of temperature. To obtain a basic representation, we consider the ideal case of a linear (unbranched) polymer chain of homogeneous chemical composition in a single phase bulk material which is amorphous (non-crystalline). For all such materials, when modulus is plotted

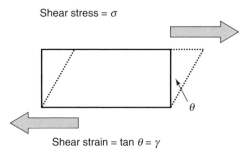

Figure 4.2 *Shear stress, strain and modulus.*

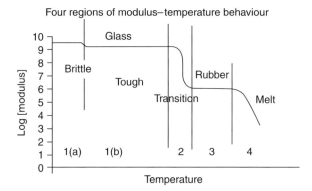

Figure 4.3 *Modulus dependence on temperature for an ideal, linear, amorphous polymer.*

against absolute temperature, a curve is obtained with four principal characteristic regions (Figure 4.3).

These are:

(1a) A high modulus region where the polymer exhibits the behaviour of a *glass*. At very low temperatures the plastic will be brittle and will fracture easily. A piece of rubber tubing placed in liquid nitrogen and subjected to a sharp impact will shatter like glass. All polymers will break easily at very low temperatures and undergo *brittle fracture*.

(1b) A region in which the modulus undergoes a small decrease. The polymer becomes more resistant to shock, although still retaining a high modulus. If now subjected to a sharp impact, the plastic tends to tear and resist fracture. This is called *ductile fracture*.

(2) A transition region where the modulus drops sharply to a plateau value where large deformations become possible with small stresses. The material has been transformed from being a glass to a rubber and this characteristic is unique to polymeric materials.

(3) An extended temperature range within which the material behaves as a *rubber*. For a thermoplastic polymer, the extent to which this rubbery behaviour exists depends on the molar mass. In the case of a thermoset polymer, the flexibility will depend on the degree of cross-linking of the polymer chains.

(4) A further transition to a viscous liquid *melt* state. In the case of thermoset polymers, this drop with increasing temperature is not observed and the modulus stays at its rubbery value until the polymer decomposes.

The small transition, which often exists at temperatures below that of the main glass to rubber transition, and which is associated with a change from a brittle to a tough glass, will be discussed in more detail in the next chapter. These four temperature regions vary along the temperature scale for different polymers.

Following convention, we use a compound adjective to describe the behaviour, noting that the overall range of properties covers both elastic (solid) and viscous (liquid) ranges. Combining these two, we use the word "*viscoelastic*". In other words, all polymers exhibit both elastic and viscous properties, with the relative importance of each determined by the temperature. Of particular importance is whether the transition region is above or below room temperature and whether the melt region is within a practicable processing range.

4.4 Working temperature range for a polymer

The working range of a plastic depends on its application, but is usually defined in terms of the mechanical properties discussed above. The modulus, expansion coefficient, density, etc. all vary with temperature in a way which is different from that observed with ceramics and metals. For load bearing and structural applications a high modulus is desirable. However, when the polymers are used in packaging, as rubbers or as sealants, greater flexibility is desirable. Plastics have the advantage of being light and relatively easy to process and as such are attractive for the fabrication of a variety of different objects. Identification of the temperature at which a polymer changes from being rigid to being flexible is critical in determining the working temperature range of that material. If the material is being used for its load bearing characteristics then this temperature is the upper working temperature; alternatively, if the material is used for its elastomeric properties then it is the lower working temperature. In practice, engineers often wish to change the working range of a plastic and several ways of changing the physical characteristics have been explored. As we shall see later, similar definitions can be created to cover the operational range for a plastic used for its vibrational damping, sound absorption, gas barrier or electrical characteristics.

For hard glassy materials the working range is often determined by the temperature spread between the brittle to tough transition and the main glass to rubber transition.

So, since the end use of a polymeric material depends critically on the characteristics of the glass to rubber transition, we deal first with this phenomenon.

4.5 The transition region

The glass to rubber transition region is determined by the onset of conformational change involving internal rotation of the polymer main chain. The temperature midpoint of the change is called the *glass transition temperature, T_g*. There are several glass transition theories, but the one most closely related to the molecular motion approach is the kinetic theory.

The glass to rubber transition occurs over a temperature range when the modulus drops from the high value characteristic of a glass to the unusually low (for a solid) value of a rubber. Typically, the modulus for a glassy polymer is of the order of 10^9 Pa and in the glass transition region it falls to a value of the order of 10^6 Pa. In this temperature range, the internal rotation motion, which permits large-scale deformation of the polymer molecule and which is absent in the glass, becomes evident in the rubber. There are two effects a rise in temperature can give to a molecule that enable this particular type of molecular motion. The first is *thermal energy*. The second, made available through thermal expansion, is an increase in the empty space between the molecules, called the *free volume*. We shall deal first with the effect of thermal energy, returning to the question of free volume later in this chapter.

4.6 Temperature and energy

Increasing the temperature of a material increases the thermal energy of the molecules. It is usual to define the thermal energy in terms of a value kT per molecule, where k is Boltzmann's constant. The corresponding value per mole is, of course, RT, where R is the gas constant. The energy will activate a number of forms of vibration, partial bond rotation as well as the complete rotation of segments of the backbone chain. It is convenient to consider the energy per rotating unit, even although the moving element is only a small part of the total molecule. However, again we can say that the thermal energy available for rotation of each such unit is kT, the change from molecules to rotating chain sections being accommodated in the units of k (energy per rotating unit). Below the transition temperature the thermal energy kT is not sufficient to power the internal rotation, whereas above the transition it is adequate for the rotation. So the transition is associated with a "critical" value of kT. This critical value is the activation energy, $\Delta E^{\ddagger}$, described in detail in Chapter 3. In terms of the kinetic theory, this process is termed "thermally activated" and may very well obey a simple Arrhenius temperature relationship. For low molar mass species the internal rotation process is accurately described by a simple Arrhenius dependence. However, the process becomes more complex as the chain length is increased. Nevertheless, the simple theory does indicate that the modulus has a value of that of a glass when $kT < \Delta E^{\ddagger}$, and a value of a rubber when $kT > \Delta E^{\ddagger}$. If we assume that the process is purely a thermally activated process, a measure of the transition temperature is $T_g = \Delta E^{\ddagger}/k$.

4.7 Electrostatic forces opposing rotation

Non-bonding interactions, associated with steric hindrance, are not the only type of force that can create an energy barrier to rotation. Also of significance are

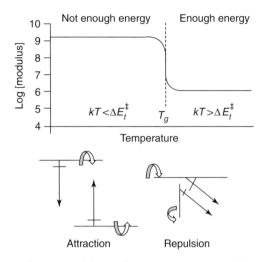

Figure 4.4 *Adequacy of thermal energy to overcome barriers to rotation.*

electrostatic forces of attraction and repulsion. These become significant when each rotating unit contains a dipole, although quadrupole and induced dipole forces can also play a part. The unit will always attempt to adopt the lowest energy state, minimising repulsions and maximising attractions. However, internal rotation will pull the dipoles out of their low energy arrangement into a higher energy, more repulsive, situation. The lowest energy state for two dipoles is when they are "paired" so that the charges in the bonds cancel one another out (Figure 4.4).

4.8 The effect of chemical composition

The temperature of the glass to rubber transition depends on the energy barriers opposing internal rotation. It is now a simple matter to determine how these depend on chemical composition.

First, from the point of view of steric interference, those chains with the largest side group substituents will have the most steric hindrance, the highest energy barriers, and so the highest transition temperatures. Consider the following three carbon–carbon backbone chains: polyethylene with only hydrogen atoms, polystyrene with one phenyl group on each monomer unit, and poly(9-vinylcarbazole), often denoted as PNVC for poly(*N*-vinylcarbazole), with a three-ring substituent on each monomer unit.

Polyethylene $\{-CH_2CH_2-\}_n$ $T_g = -60\ °C$

Polystyrene {−CH$_2$CH(C$_6$H$_5$)−}$_n$ T_g=100 °C

Poly(9-vinylcarbazole) {−CH$_2$CH(C$_{12}$H$_8$N)−}n T_g=220 °C

In the same way, we can compare chains of similar steric volume but different polarity. Consider non-polar polypropylene and polar poly(vinyl chloride) (PVC).

Polypropylene −CH$_2$CH(CH$_3$)− T_g=−20 °C (atactic)

Poly(vinyl chloride) −CH$_2$CHCl− T_g=80 °C

The polyethylene chain, with only hydrogen atoms, has the lowest steric hindrance, the lowest energy barrier and the lowest T_g (Figure 4.5). Polystyrene has more steric hindrance and so a higher barrier, requiring higher thermal energy, giving a higher transition temperature. The three-ring carbazolyl substituents create the most steric interference, the highest energy barrier, so require the highest thermal energy, giving the highest transition temperature (Figure 4.5).

The examples of polypropylene and of poly(vinyl chloride) require some further comment. After all, much of the poly(vinyl chloride) that we use at room temperature is flexible and "leather-like", while commercial polypropylene is stiff. These observations are directly contrary to the glass transition temperatures listed above.

However, the four regions of mechanical behaviour that have been described, and the transition temperatures listed, are for the ideal case of a linear, amorphous, unadulterated polymer. Neither of these two satisfy these "ideality" conditions (Figure 4.6).

As will be discussed in Chapter 6, commercial polypropylene is isotactic and crystalline at room temperature. Thus the modulus is that of a partially crystalline

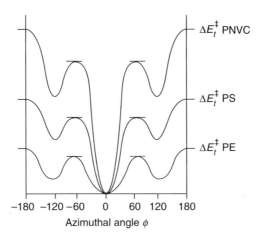

Figure 4.5 *Comparison of polyethylene, polystyrene and poly(9-vinylcarbazole).*

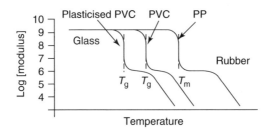

Figure 4.6 *Crystalline and plasticised polymers.*

material and not that of the softer amorphous atactic polymer. Commercial poly(vinyl chloride) is softened by the addition of a plasticiser.

The mechanism of plasticisation of PVC is interesting. We have seen that the high glass transition is a result of electrostatic forces between carbon–chlorine dipoles opposing chain internal rotations. Consequently, anything that will screen or diminish the interdipole forces, and indeed the interchain separation, will thereby lower the transition temperature. This is exactly what the plasticiser molecules do.

Firstly, they have to be at least partly soluble in the polymer, which requires them to be moderately polar. Secondly, they must have non-polar, very flexible groups to allow molecular movement. A typical plasticiser is dioctyl phthalate.

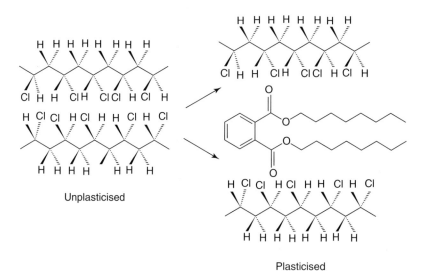

Figure 4.7 *Plasticisation of a polar polymer by dioctyl phthalate.*

The carbonyl dipoles of the ester groups are attracted to the carbon–chlorine dipoles of the chain, whereupon the non-polar octyl groups separate those dipoles from those on other chain segments (Figure 4.7).

In order for the plasticiser molecule to dissolve in the polymer, it often has a low molecular weight. As a result, not only can it go in, over time it can also diffuse and leach out. This is a very significant problem with many uses of PVC as it leaves the polymer hard and brittle, and coats the surface with a liquid that may be toxic. The leaching of plasticiser from various plastic components also used to give cars the "new car smell". This problem of plasticiser "blooming" is solved by the incorporation of low molecular mass polyester. This produces the expansion of the lattice but does not migrate to the surface as quickly because it is interacting with the chain through many functional groups.

4.9 The effect of time

We can summarise the foregoing by two questions and answers.

First, "What are the molecules doing?" Answer: "The molecules are moving."

Then, "Why do the molecules move?" Answer: "The molecules have an excess of energy."

We now move on to a third question: "What else do the molecules need to allow them to move?"

Molecules and segments of polymers, unlike photons or electrons, require a finite amount of time to move. Electrons and photons move so quickly that in the timescale of most mechanical experiments and observations, the movement

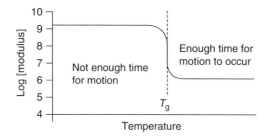

Figure 4.8 *Observation time and deformation time.*

is instantaneous. On the other hand, the diffusion of molecules and the conformation changes of large sections of macromolecules are relatively slow, requiring times that may be significant within the timescale of observation or use. In order that an observer can witness the deformation of a sample with the characteristics of rubber, the observation time must be long enough to allow the necessary shape change of the molecules. In other words, the answer to our question is: "The molecules need time!"

This now allows us to characterise the glass to rubber transition as that change observed when the molecules move at such a speed that the internal rotation can take place within the time associated with the observation. In other words, in the glass the speed of molecular motion is too slow for any significant change to be observed, whereas in the rubber the molecules move sufficiently rapidly that gross deformation is observable. Now, time and temperature are related since conformational changes are thermally activated processes and increasing the temperature speeds up molecular motion (Figure 4.8).

For a fixed time of observation, the increase in temperature in the glass speeds up the molecular rotation, decreasing the required time. So, from the point of view of time, we can define the T_g as that temperature when the time required for molecular motion equals the time available in the observation. The time to make an observation will vary with the type of experiment which is being performed.

We can say the same thing in terms of reciprocal times, or rates. The transition occurs when the rate of molecular internal rotation and the macroscopic rate of stress–strain application are equal.

In other words, in order to show the properties of a rubber, the molecules need both energy and time. In Chapter 3 we indicated that the motion of an element of a polymer chain involves the cooperative movement of a group of monomers and so requires more energy than that associated with a single bond rotation. So it is not surprising that the time required for this to occur is much longer than that for rotation round a single bond.

We come now to a very important corollary of this time requirement. Different experiments, different uses and different observations will all involve different times. Consequently, an observer will "see" the transitions whenever different rates of internal rotation equal the different rates of observation. In turn, the different rates of internal rotation will occur at different temperatures, and so the observed transition temperature will depend on the time of observation, that is on the time of an experiment or use. This is so important that it is worth repeating:

The observed transition temperature is a function of the time of observation.

The simple concept of the time–temperature dependence of the process makes the glass to rubber transition intrinsically different from a process such as melting. Melting is essentially defined by the break up of the forces which hold the molecules in the solid state and occurs at a well defined temperature which is defined by the enthalpy of melting. The melting point does not vary with the method of measurement and any small differences observed are a consequence of the time it takes for the energy to be conducted through the sample. A surprising amount of confusion exists in the literature on the glass transition because of a lack of recognition of this time–temperature duality in the transition process.

Let us look at the modulus–temperature relationship with measurements made in two very different times. First, we reproduce our familiar diagram making use of measurements made over long times. Then the transition occurs when the molecules can move in these longer times, a slow movement that can be achieved at low temperatures.

However, in the second set of measurements, the observations are made over very short times. Consequently, the transition is seen only when the molecules are moving very rapidly, a phenomenon that requires high temperatures (Figure 4.9).

The rule is: for the transition on the temperature axis, long observation times "see" slow molecules and so correspond to low transition temperatures; short observation times "see" only fast molecules and so correspond to high transition temperatures.

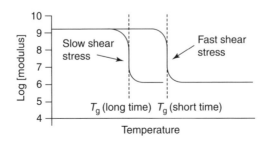

Figure 4.9   *Observed transition in long time and short time measurements.*

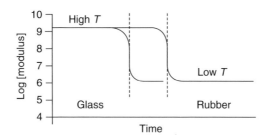

Figure 4.10 *Modulus dependence on time at high and low temperatures.*

We are now able to consider modulus as a function of time at any given temperature (Figure 4.10).

An observation made at low temperatures will only "see" the movement of slow molecules after long times. Conversely, an observation at high temperatures will find the molecules moving rapidly, and so will "see" the transition at short times.

Another important distinction between the transition and the melt process is the breadth of temperature over which the process is observed. In the case of a melting process, the transformation from a solid to a free flowing melt will occur over a temperature interval of around 1 K or less. So it is possible to use the melting point of a solid as a method of identification of the material and its purity. As indicated in Figure 4.3, the change of the modulus from that of a rigid solid to an elastomer will typically occur over around 10 K, reflecting the "relaxation" nature for the rotational isomeric process. This will be discussed in more detail in Chapter 10. So a broad transition is a characteristic of the glass to rubber transition.

4.10 The equivalence of time and temperature

In the glass to rubber transition, time and temperature are related through the speed of movement of the molecules undergoing internal rotation. We now have to quantify that relationship. The key is that the transition is observed when the rates (times) of the molecular rotation process and those of the observation are equal.

Starting with the molecular movement, the rate of a molecular process crossing an activation energy barrier can be related to temperature through the Arrhenius equation.

$$\text{Rate of molecular process} = A\exp(-\Delta E^{\ddagger}/RT)$$

where A is a constant and $\Delta E^{\ddagger}$ is the activation energy in units per mole. Very often the exponential is written as $-\Delta E^{\ddagger}/kT$, where k is Boltzmann's constant and the energy is in units per molecule.

Then if the molecular and macroscopic rates (times) are equal at the transition, the molecular rate can be replaced by the observation rate or reciprocal time and, of course, T must be replaced by T_g. Then

$$\text{Rate of stress/strain} = \text{reciprocal observation time} = A\exp(-\Delta E_r^\ddagger / RT_g)$$

where $-\Delta E_r^\ddagger$ is the activation energy per mole for rotation. This is a most important result since it allows us to relate modulus, time and temperature in a single relationship. This can be illustrated in the familiar Arrhenius diagram. To construct the diagram we first plot the logarithm of the rate of internal rotation against reciprocal absolute temperature, and then simply relabel the axes to logarithm of rate of macroscopic process and reciprocal transition temperature. Then the graph is a straight line of slope $-\Delta E_r^\ddagger / R$.

Very significantly, we can use this diagram to ascertain how the transition temperature changes with the speed at which the observation of a sample is being made and will vary with the different rotational energy barriers. First we plot the ln rate against $1/T$ graph for a polymer with a large rotational energy barrier, and then we do the same for a polymer with a low rotational energy barrier. Not unexpectedly, the rotation crossing the low barrier is faster, and occurs at lower temperatures, than the rotation impeded by a high barrier.

Next we select two rates, now macroscopic observation rates, one high and the other low. For these we read off the reciprocal transition temperature for each polymer at each rate and examine how this changes (Figure 4.11).

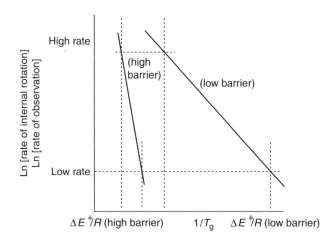

Figure 4.11 *Arrhenius diagram relating rate of observation to transition temperature for polymers of high and low rotational energy barriers.*

The convergence of the lines carries a significant, and often not appreciated, consequence. At the highest rates, as the reciprocal transition temperatures approach equality, so too must the transition temperatures themselves. This means that at these high rates, the transition temperature changes most for the polymers with low rotational energy barriers. In other words, the transition temperatures at high rates of stress and strain are most sensitive to time in polymers that are rubbers at room temperature, and are less sensitive to time in polymers that are hard glasses at room temperature.

This is important in rubber technology because it means that a rubber that is above its transition temperature in some low rate, long time use could be below the transition temperature, and so become a breakable glass in some high rate, very short time use.

4.11   Non-Arrhenius activation

There are many assumptions inherent in the use of the Arrhenius equation. The most significant of these is that the molecules or parts of molecules under consideration are not affected by other molecules or parts of molecules. This is clearly not the case when we consider the change of shape of very large molecules in a solid polymer.

Actually, for each bond the probability of the rotation moving through the high energy barrier of the eclipsed state at 180° is small compared to that of the partial rotation from 0° to 120°. Under these conditions the chain never really undergoes totally free rotation about the backbone, but twists and follows a contour which avoids the highest energy form but still executes the required movement in space. If several bonds undergo a collective motion, as was indicated in Chapter 3, the "180° movement" can be achieved without having to invoke a transition through the single bond 180° barrier. In fact the glass transition movement is usually the cooperative motion of eight or more bonds. So the activation energy which is measured experimentally by observing the rate/temperature dependence is really a composite value for a number of cooperative movements and is dominated by the lower energy processes. Then the profile for the rotation of one bond in Figure 4.5 must be seen as a symbolic guide to the effective composite rotational profile.

The Arrhenius equation predicts that no matter how low the temperature, internal rotation can always be "seen" provided that the observer is prepared to wait a long enough time. In other words, the straight line in the Arrhenius diagram can be extrapolated to infinitely low temperatures and infinitely long times. In fact this is not the case. It is found that there is always a minimum transition temperature below which molecular internal rotation can never be observed,

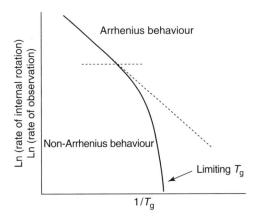

Figure 4.12 *Limiting long time transition temperature.*

no matter how long we wait. On the Arrhenius diagram this shows as a curvature to an asymptotic minimum T_g (or to a maximum in its reciprocal) (Figure 4.12).

Several explanations have been put forward for this non-Arrhenius curvature to a limiting minimum transition temperature. However, the simplest is to return to the two properties introduced in section 4.5, and note that decreasing the temperature decreases the volume of a material and so decreases the empty space between the molecules in the sample. We assume that for a rotating unit to move there must be a certain amount of empty space, called free volume, between the rotating unit and its neighbours. If this empty space falls below a critical value then movement becomes impossible. When this becomes important, the movement of the molecule is governed by the availability of free volume rather than just by its internal energy. This concept was first developed to explain diffusion in liquids, and then extended to explain transitions in polymers. The equations derived on this basis indeed predict the curvature observed in the logarithm rate against $1/T_g$ Arrhenius diagram.

The availability of space is influenced by similar factors to those which influence the internal rotational barrier, i.e. the non-bonding interactions. However, in the case of free volume, it is the strength of the polymer–polymer chain interactions which is important. The energy which thermally activates the internal rotation can also increase the separation of the polymer chains, so these two effects are coupled and difficult to separate.

Since the side group movement involves just a conformational change of a pendant group, it will not require additional volume and so more closely exhibits the characteristics of a simple thermally activated process. Likewise the libration movement (discussed in Chapter 5), which starts at temperatures below the main glass to rubber transition, has smaller volume requirements, and so again more closely follows the Arrhenius equation.

The conclusion to be drawn from the discussion above is that the high, and varying, slope of the Arrhenius diagram at low rates and transition temperatures is due to the free volume requirement, while the smaller slope at higher rates and temperatures reflects the cooperative nature of the overall conformation change.

4.12 Free volume and activation energy for movement in the glass

In the glass, there is insufficient free volume for the chains to be able to move. What then defines the critical size of the free volume that is required for backbone motion?

This movement requires the cooperative motion of a number of bonds and the free volume requirement is reflected in the energy requirement. The activation energy for single bond rotation has typical values of the order of $10–40\,kJ\,mol^{-1}$, whereas typical values for the glass to rubber process will be in the range $100–300\,kJ\,mol^{-1}$. These values are so high that when measurements were first made it was suggested that rotation of the elements of the backbone might involve breaking and reforming of chemical bonds. This idea is incorrect, but it does indicate that the energy required for backbone rotation is much the same as that for bond breaking and re-formation. The explanation is fairly simple; the backbone rotation process does not occur about a single chemical bond, but involves typically between six and eight bonds. The precise number involved will depend on the chemistry of the polymer. It is easy to see that if the activation energy for a single bond rotation is multiplied by the number of bonds moving, the observed values are easily achieved. This concerted movement is a little like the inversion of the cyclohexane ring, where several bonds move together to achieve the boat to chair conformation change. If the motion is to occur without translation then it has to have the form introduced in Figure 3.12.

In order for some of the *trans* conformations to change into *gauche* and vice versa, but not to undergo translation, a finite space is required for the cooperative process. This is the free volume required for the backbone rotation that changes the glass into a rubber.

A very clear indication of the importance of free volume is the way that the observed transition temperature changes with pressure. The application of pressure decreases the free volume and so requires further thermal expansion to enable the chain movement. This, of course, raises the observed transition temperature. This is important when one looks at polymers used in applications where the material is subjected to high pressure, such as seals in diving bells, hydraulic systems, etc. A polymer at atmospheric pressure may be above its transition temperature and so its elastomeric properties allow it to provide an effective seal. However, application of pressure, such as that experienced when a diving bell is lowered into the depths of the sea, may be sufficient for the material to be forced below its transition temperature, become glassy and cease to be able to act as a seal. If the material is being used to seal the windows, this could have disastrous consequences.

So now we can extend the answer to our question: "What do the molecules need?" Answer: "The molecules need energy, time and space."

4.13 Volume and time

The consideration of free space changing with temperature brings us to some of the earliest studies of the effect of time on both the transition temperature and the polymer volume. A French scientist, Kovacs, carried these out. He heated a sample of polymer to the rubber state and then cooled it to the glassy state, all the while making precise measurements of the volume (Figure 4.13).

Polymers, like most other solids, exhibit a simple linear expansion coefficient in the glassy state. The expansion coefficient reflects the effect of the increasing thermal energy increasing the chain to chain distance. The striking difference between a simple organic solid and a polymer is that the expansion, measured as volume against temperature, changes its slope at some temperature and then often continues in an approximately linear manner up to the melting point. The thermal expansion coefficient has changed at the glass to rubber transition. Then, once again, the longer times associated with slow cooling yield a lower transition temperature than the shorter times of fast cooling.

An important observation in the figure is that the volume, and so the density, of the glass is not constant at a given temperature, but is a function of the cooling rate! Thus a polymer glass is not a solid in thermodynamic equilibrium, but is, in a sense, "supercooled" to various states of non-equilibrium molecular packing. If the polymer is quenched from a high temperature, more entropy is frozen into

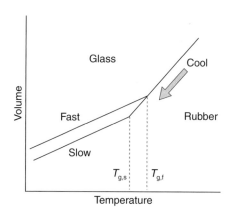

Figure 4.13 *Thermal contraction of a polymer on cooling.*

the material and as a consequence the T_g that is observed depends on both the rate of cooling and the temperature from which the cooling started.

This dependence on cooling rate is not unique to volume and density. It will be shown, to a greater or lesser extent, by all the thermodynamic properties of the glass, and so it will become apparent in a variety of different thermal analyses. A lack of recognition of these rate effects leads to different measurements that may appear to be inconsistent. Thus often the thermal mass of a sample used in mechanical measurements is much larger than that used in calorimetric investigations, so the rates of heat flow and temperature rise are different.

This concept of free volume is not a very convenient way of quantifying the restriction of the motion, although it does reflect that nature of the changes which occur at a molecular level. Nevertheless, there are techniques to probe the volume between molecules and it is found that, as assumed by the free volume theories, the available space between molecules does dramatically increase at the transition temperature.

4.14 Other glass transition theories

All the foregoing discussion is a presentation of the kinetic theory of the glass transition. It has many merits, not least being that it is the theory most easily understood by persons with a background in a molecular science. However, it is not the only theory, and we should be aware of the existence of alternative explanations of the processes that we have just described.

When discussing the limiting transition temperature at long observation times, we mentioned the free volume theory of the transition. This had its origins in a treatment of diffusion in liquids put forward by Doolittle. In it, molecules diffuse by jumping into a "hole" that opens up between them and their neighbours. This theory is based on consideration of the fluctuation in the intermolecular free volume. The theory was extended to polymer molecules by Bueche and by Meares, and expressions were derived relating the probability of molecular movement to the critical free volume required and the actual free volume available at any given temperature. As mentioned above, the derived equations give a curved dependence of logarithm of rate on reciprocal temperature, as is found in experiments. So the theory provides a good description of the observed behaviour at low temperatures and long times.

A quite different approach is adopted in the statistical mechanical theory expounded by Gibbs and DiMarzio. They considered that the rotating unit could exist in two stable conformations (which might correspond to the *trans* and *gauche* conformers discussed earlier), separated by a standard state energy difference, $\Delta E°$, called the *flex energy*. Using the techniques of statistical thermodynamics, they calculated the partition of the units between these two forms. In theory, the

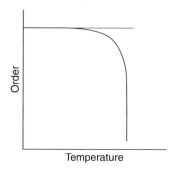

Figure 4.14 *Order–disorder transition.*

population of the lower energy form should increase as the temperature is lowered, so that the disorder in the arrangement between forms also reduces as the temperature is lowered. The phenomenon is rather similar to the order–disorder transitions familiar in metal alloys (Figure 4.14).

The asymptotic approach to a limiting low temperature value is apparent. However, in its simple form, the theory assumes that this is an equilibrium value of the distribution and order, whereas we know that the glass is a non-equilibrium state. The theory has been extended using non-equilibrium thermodynamics, but is not as useful in technological applications as is the simple kinetic theory elaborated above.

4.15 Time, rate and frequency

In the treatment above, we have dealt with time and its reciprocal, rate. The concept of "rate of molecular rotation" is clear enough. However, we were not specific about what we meant by "rate of observation". We can look at this in two ways.

The first is to consider the time over which observations are made, and then simply take its reciprocal. For example, in the volume contraction experiments of Kovacs, "rate of cooling" is measured as degrees per unit time.

However, an alternative concept arises if the action on the polymer is carried out in a periodic, cyclical way. In this section we are concerned with mechanical properties, so we could apply a deforming stress in an alternating push–pull fashion. The resulting strain would also be periodic, being extension during the pull and contraction on the push. Then the most convenient measurement of the time or rate of the stress/strain observation is the frequency of the applied stress. High frequencies correspond to stress applied over short times, while low frequencies correspond to stress applied over long times. In this way the "time of the experiment" as experienced by the molecules is the time of one cycle of the periodic stress/strain, and not the total time over

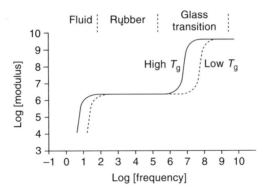

Figure 4.15 *Modulus transitions with frequency at different temperatures.*

which we can measure the effect. In other words, by applying a high frequency stress (one cycle of which is far too fast for the unaided eye to see) over a prolonged period of time, we can observe the effects in a perfectly measurable experiment.

Consequently, the modulus, being a function of time, is also a function of frequency (Figure 4.15).

The modulus–logarithm frequency curve is a mirror image of the modulus–logarithm time curve. At higher temperatures, the molecules are moving faster, and so the transitions are moved to higher frequencies (broken line).

4.16 The Williams–Landel–Ferry equation

Long before scientists had developed a full treatment of the molecular processes underlying the glass to rubber transition, engineers and technologists needed a formula to convert stress/strain/modulus data measured at one temperature to data at other temperatures. What was wanted was not an involved equation with many components, but a simple adjustment factor such as multiplication or division.

Williams, Landel and Ferry noted that, since time and temperature had an equivalent effect, adjusting the time could accommodate the result of changing the temperature. In other words, for a modulus which was a function of temperature T_1 and time t_1, a new modulus for temperature T_2 could be given by:

$$E(T_1, t_1) = E(T_2, t_1/a_T)$$

where a_T is a simple multiplier/divider containing only one variable (temperature). It does, though, contain three constants that are specific to the polymer under investigation. These authors showed that this adjustment factor could be given by a totally empirical formula, which is now called the

Williams–Landel–Ferry (WLF) equation:

$$-\log(a_T) = \frac{C_1\left(T - T_0\right)}{C_2 + \left(T - T_0\right)}$$

where T is the variable leading to the new temperature T_2, T_0 is constant and is a reference temperature for the polymer under study, while C_1 and C_2 are constants, again characteristic of the polymer under study. Since the equation was first put forward, many workers have evaluated these three constants for almost all polymers in industrial use, and they are available in many books and compendia of polymer characteristics.

However, some comments on this equation are appropriate. First, although many people have attempted to rationalise it in terms of other polymer properties, it is in fact a totally empirical relationship with no underlying scientific reason for its form. Its justification lies in its great usefulness, not in its molecular significance. Secondly, the fact that it gives a good fit to experimental observations is hardly surprising, since a three-constant equation can usually provide a good fit to experimental data. That said, it is rather surprising that a single multiplying factor so well accommodates the time–temperature equivalence of any given polymer.

It turns out that, for a given polymer, the observed reference temperature, T_0, is rather close to the limiting glass transition temperature measured at the longest times (lowest rates). The constants C_1 (dimensionless) and C_2 (absolute temperature) are different for different polymers. However, for polymers with similar glass transition temperatures, the slopes of the Arrhenius plots are almost the same, meaning that the relationships between logarithm rate and temperature are almost the same, and so the values of the WLF constants are so close to each other that average values can be used for this group of polymer types. Since many commercial thermoplastics must be glass-like at room temperature, but be able to be processed at temperatures not too far above 120 °C, they have similar transition temperatures around 100 °C, and so have similar WLF constants, average values being 17.4 for C_1 and 51.6 K for C_2. PVC (T_g = 80 °C), polystyrene (T_g = 100 °C), poly(methyl methacrylate) (T_g = 100 °C) are typical examples. In the past, this has led many reports and books to state that the constants are "universal" for all polymers, which for all possible polymer types they are not.

Let us now look at the use of this equation. Suppose that stress/strain measurements are conducted at a temperature T_1 and at a variety of stressing rates $r_1 < r_2 < r_3$, etc. (Figure 4.16).

If this is repeated at different temperatures $T_1 < T_2 < T_3$, etc., and we now move to logarithm (time = reciprocal rate) to convert multiplication by a_T to

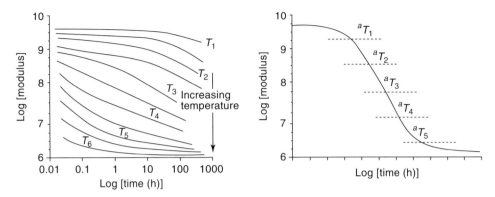

Figure 4.16 *Stress/strain curves at different strain rates; moduli moved along the rate axis to superimpose data at different temperatures.*

addition, then we find that a "master curve" results when we move the data at each temperature along the logarithm time axis by the factor a_T (Figure 4.16). So the amount of movement required to superimpose moduli at any temperature on the master curve can be used to measure a_T. Alternatively, a_T can be found in a book and used to calculate the amount of movement required to find the modulus–time curve at a new temperature (Figure 4.17).

4.17 Softening and melting

Although the polymer glass is disordered, the solid is held together by forces which are similar to those which define the melting temperature of a simple solid. Van der Waals, dipole to dipole, and dipole to induced dipole forces bind the elements of the chain to one another. Surprisingly, it is possible to estimate the temperature at which a polymer will melt to a liquid by adding together the components of the overall interactions which exist in the solid state. This only

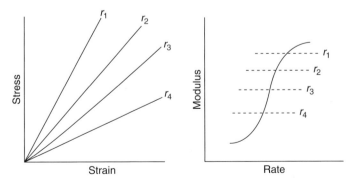

Figure 4.17 *Stress, strain and modulus at different strain rates.*

requires knowledge of the chemical nature of the repeat structure of the polymer backbone. These simple *additivity* calculations are surprisingly useful and are the basis of computer predictions for amorphous polymers. Further, if we use the bond additivity relationships to calculate the "flow melting points" of the polymers, T_m, these exhibit a simple relationship to the glass transition temperatures:

$$T_g = \sim 0.65 T_m$$

So there is a close correlation between the glass to rubber, and the melt to liquid processes. The connection is that for both processes the lattice has to be expanded to a point at which the polymer molecules are sufficiently separated to allow large sections of the chain to move, first only by conformation change and then by reptation, a type of motion described in more detail in Chapter 8.

4.18 Summary

The transition glass to rubber is observed when the internal rotation and shape change of the molecules can take place during the time of observation. For this to happen the molecules must have energy (to overcome the barriers to rotation), time (to rotate during observation) and space (into which a moving segment of a molecule can rotate). As a result, the transition is observed as a function of temperature and also time or frequency. There are a number of different methods of observing the glass transition process and each reveals a different facet of the molecular processes discussed above.

Further reading

Turi E.A. *Thermal Characterization of Polymeric Materials*, 2nd edn., Academic Press, San Diego, 1998.

5

The glass state

In this chapter we examine the properties of polymeric materials in the glass state, the lowest of the four temperature regions.

5.1   Polymer glasses

In the foregoing we have described the glassy state as being a very disordered state of high modulus with neither complete internal rotational nor translational motions taking place. We now look at this in more detail.

When we examine the modulus–temperature curve of a glass, and again consider the ideal case of a linear amorphous polymer, we find that the relationship is not a straight line, but has minor transitions at temperatures below that of the main glass to rubber transition (Figure 5.1).

The transitions are named in order of decreasing temperature, starting with the main glass to rubber transition, using the Greek alphabet, α, β, γ, δ, ε, etc. Since the transitions occur at progressively lower temperatures, they are associated with the onset of molecular processes with progressively lower energy barriers.

5.2   Low-temperature molecular motion

If full 360° internal rotation is not possible, what then are the motions that give rise to these subsidiary transitions? There are two of importance.

The first of these is partial rotation to some state higher in energy than the lowest energy state, but still lower in energy than the energy barrier of the eclipsed state. One such state might be the *gauche* (cisoid) state discussed in Chapter 2. Such a partial rotation to and fro about the lowest energy point is called *libration* (Figure 5.2). Libration motion can be considered as motion which is restricted to the lower part of the potential well or as cisoid transitions which are restricted to the movement of groups which involve only one or two bonds. The ends of polymer chains, being sufficiently unhindered, allow such motions to occur rather easily. However, inside the chains these motions are restricted as they require movement of other parts of the molecule.

A second molecular motion that might be able to take place at low temperatures is rotation of, or inside, the substituent side group on the chain. Very often the libration and *side group rotation* are coupled, so that both start up together

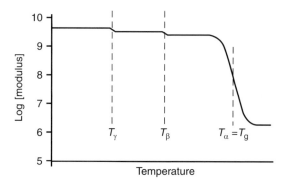

Figure 5.1   *Transitions occurring in the glassy phase.*

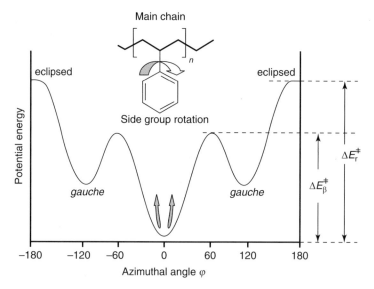

Figure 5.2   *Energetics of librational motion.*

at the same temperatures and times. The common feature of both types of motion is that they can occur without the requirement for translational motion of the main backbone chain and hence do not need an activation volume. Consequently, the activation energy is independent of pressure, application of which does not change the temperature of transition.

5.3   The effect of libration and low-temperature motions

The small amount of deformation made possible by the libration, and evidenced in the β-transition, has a very important technological consequence. The very complex mechanism of fracture is treated in Chapter 9. Here we simply say

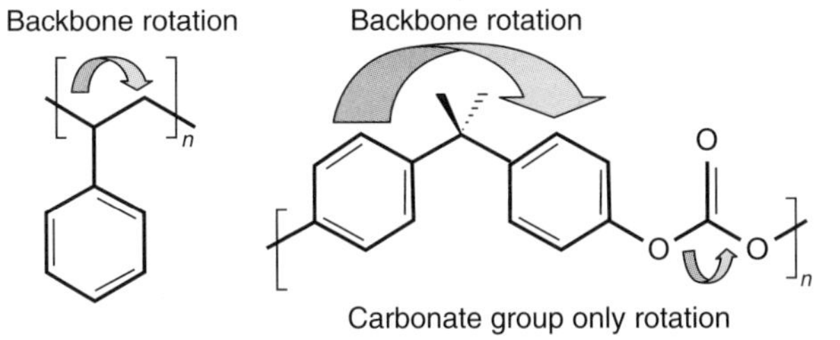

Figure 5.3 *Segmental rotation in polystyrene and polycarbonate.*

that the movement allows the polymer to absorb an impact without shattering. In other words, in the temperature region between the main glass transition and the β-transition, we have a *tough* glass. An analogy is the way a world champion boxer "rides" a punch on the jaw by swaying his head back with the punch. On the other hand, a less skilled boxer, who cannot move his head out of the way, receives the full force of the punch, with disastrous effects.

To see the effect of this in technology, consider the difference between polystyrene and polycarbonate. In polystyrene, the β-transition is above room temperature and the plastic is brittle, cracking and shattering very easily. On the other hand, in polycarbonate, the glass transition is well above room temperature, but the β-transition is well below room temperature. We have, then, a very tough glass useful for many different impact resisting applications, including things like shatter-proof glass, protection against vandalism, and even resistance to low velocity bullets. The molecular difference between the two polymers is sketched in Figure 5.3.

In polystyrene, the energies required for libration and for full rotation are rather similar. As a result, the two transitions are close in temperature, and both are above room temperature. However, in polycarbonate, the relatively unhindered carbonate group can rotate easily, with a low energy barrier, while the rotation of the whole bisphenol A bicarbonate unit is difficult, and has to overcome large steric hindrance. As a result, only slight thermal energy is required for the smaller β-process, while a large energy is required for the full α-process. The two transitions are, therefore, well separated on the temperature scale, with room temperature lying conveniently almost midway between them.

5.4 The effect of time on the α- and β-transitions

Having looked at the effect of temperature, now we must consider the effect of observation time or rate. This is done in the same way as with the effect of chemical composition on the glass transition. The β-transition is caused by the onset of molecular motion of lower energy barrier than that involved in the full rotation

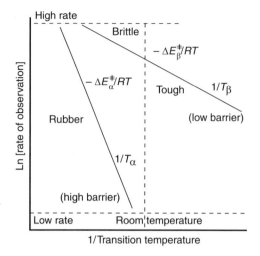

Figure 5.4 *Convergence of transition temperatures at high rates.*

causing the glass transition. So again the Arrhenius diagram is constructed, plotting the natural logarithm of the rate of the molecular process against reciprocal temperature. Since the rate of the molecular motion equals the rate of the experiment when T is the transition temperature, the experimental rate is plotted against the reciprocal transition temperatures.

Because the libration, being of a single bond, has a smaller activation energy than does the composite motion leading to full rotation at the glass transition, the slope of the Arrhenius plot for the β-transition is less steep than that for the α (glass) transition. Thus the two lines converge at the highest rates (shortest times). Also, as before, the change in the transition temperature is greatest for the process of lowest activation energy.

This is of tremendous technological importance. Thus, considering a polymer like polycarbonate, with a glass transition well above room temperature and a β-transition well below room temperature, as the time of use is decreased, the β-transition moves towards the α-transition. In so doing it can move above room temperature. Then a glass that is considered to be tough and shock–resistant becomes brittle at high shock rates. This is the case illustrated in Figure 5.4.

5.5 Lower temperature transitions

In the foregoing, attention has been concentrated on the main glass to rubber transition and the β-transition. To these we have ascribed the molecular motions equivalent to full 360° rotation and of libration (partial rotation). Of course many other types of low energy conformation change are possible, giving rise to the γ, δ, ε and other transitions.

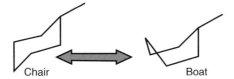

Figure 5.5   *Boat and chair forms of cyclohexyl groups.*

Side group rotations become important in polymers with ester side chains, of which there are very many. Of technological significance is the low temperature transition in poly(methyl methacrylate). Another interesting conformation change is the boat–chair switch in cyclohexyl groups (Figure 5.5).

It turns out that above the relevant transition temperature this occurs almost as easily in a solid polymer as in liquid cyclohexane.

These simple conformational processes are thermally activated and follow the simple Arrhenius equation. They are not limited by deficiencies in free volume, as the cyclohexyl case exemplifies. However, since they have a very low activation energy barrier, the Arrhenius plot has a very low slope and the transition temperature does vary significantly with the rate or timescale of the observation or use.

5.6   A glass may have different entropy characteristics

The melting point of a simple solid is a characteristic which identifies that material as it reflects the averaged interaction between the molecules present. The T_m is a first-order transition and, depending on the purity of the material, will occur over a very narrow temperature range. In contrast, the T_g has some of the characteristics of a second-order transition and is observed to occur over a broad temperature range. However, unlike an equilibrium thermodynamic second-order transition, this can depend on how the material has been treated prior to the measurement. As was seen when describing the thermal expansion coefficient in Chapter 4, if the polymer is raised to a temperature well above T_g and then quickly cooled, the amount of entropy in the glass will depend on the rate at which it is cooled. If the material is cooled infinitely slowly, the chains can reach thermodynamic equilibrium at every stage. Then, at the point at which the system changes to glass, the entropy content reflects the value at the relevant transition temperature. However, if the material is frozen very quickly then the polymer chains cannot move to adopt the minimum energy configuration and additional entropy is trapped in the system. The apparent transition temperature of the material is shifted upwards, and the entropy of the glass is that which had been in the matrix when it was quenched.

A concept closely related to the variable non-equilibrium entropy trapped in the glass is called *fragility*. It is important to realise at the outset that this particular use of the adjective "fragile" is not the same as "brittle". At first sight this seems odd, since in non-scientific everyday use the words are almost synonymous. In fact, fragility theories were developed to describe the viscosity of a supercooling liquid as the temperature is lowered to the glass transition, whereas theories of brittleness apply to crack propagation in fracture at temperatures below that of the β-transition.

The fragility concept was first introduced because the relationship between viscosity of a liquid and temperature as it supercools to form a glass deviates from the normal linear Arrhenius plot for a thermally activated process. The graph exhibits a greater or lesser degree of curvature. This deviation from the straight line is called the *fragility index*. In the first development of the theory it was applied to glass-forming liquids like glycerol, but not to polymers at the rubber to glass transition.

However, the concept was later extended to relate to the residual entropy (and so the non-equilibrium thermodynamic properties) trapped in the glass during the supercooling process. At this stage it was applied to polymers above, through, and very slightly below the glass transition temperature.

In fact, the index reflects the intermolecular interactions in the glass and is also an indication of the extent of motion of the relaxing element. Formation of polysilicate glass, like other inorganic glass-forming liquids, takes place with cooperative motions somewhat similar to those of a polymer, but only to a very limited extent and with strong interatomic forces. In these glasses the relaxing unit is analogous to the shorter relaxing elements in a polymer system. This is because the relaxing unit in an inorganic glass is restricted in its motion by the surrounding charged clusters which exist prior to vitrification. It has a low fragility index. On the other hand, polymer rubbers and glasses, where the transition requires motions which have larger amplitude and are restricted by entanglements with other chains (as we shall see in Chapter 8), usually have a high fragility index.

The fragility index can be used to estimate the amount of excess non-equilibrium entropy that can be frozen into the glass and so is useful in understanding the thermodynamic properties of a glass, whether it be an inorganic solid or an organic polymer. Generally, the higher the index, the greater the excess entropy.

The fact that plastics are frozen in a non-equilibrium state has consequences for all the physical properties that depend on density and also for the long term dimensional stability. Because the "supercooled" state is thermodynamically metastable, the polymer molecules move slowly in an attempt to reach equilibrium. These very slow molecular motions, even in the glassy state, are extensions of the processes which are observed in shorter times above the glass to rubber transition. This long time phenomenon is often termed *entropy relaxation*.

Further reading

Trant M.R. and Hill A.J. (Eds.) *Structure and Properties of Glassy Polymers, ACS Symposium Series 710,* American Chemical Society, Washington, 1998.

6

Crystallinity

In this chapter, the ordered structures that can be found in certain polymer solids are considered. As was presented in Chapter 1, the understanding of the phase morphology of a material is an essential component of "the bridges of understanding" between chemical structure and physical properties. In addition, the morphology is intimately connected with the various phenomena of molecular motion. Some consequences of this will be developed further in Chapter 9.

6.1 Ordered structure in crystalline polymers

Minerals often have strikingly regular shapes, very smooth surfaces and facets. These crystalline properties originate in the regular three-dimensional arrangements of the constituent atoms or molecules. These crystals can be grown to have dimensions of the order of many centimetres. The three-dimensional order in these materials can be confirmed by observation of their X-ray or electron scattering. The scattered radiation forms spots from large single crystals and sharp rings from powders, as shown in Figure 6.1.

Since polymer molecules are long chains, it may seem unlikely that they can form such a regular three-dimensional arrangement. However, if polymer chains can form regular helical conformations, and then these helices can come together side by side, a regular three-dimensional arrangement of polymer molecules can be achieved. So some polymeric materials show X-ray diffraction patterns with sharp rings superimposed on a diffuse background. This is shown in Figure 6.2.

This evidence suggests that there are many small regions in the bulk that have regular three-dimensional arrangements, but also regions that are not crystalline. These crystalline and non-crystalline or amorphous regions coexist, so the material is called *semi-crystalline*. In addition, the breadths of the rings found for normal semi-crystalline polymers are rather broad, indicating that the crystals in the bulk are relatively small.

6.2 Requirements for the formation of ordered structures

Not all polymers can form an ordered structure in the solid state. The requirements for a polymer to crystallise are:

- *Chemical regularity along the chain.* This is the prime factor. If chemical regularity cannot be found in the chain then it will be very unlikely that the

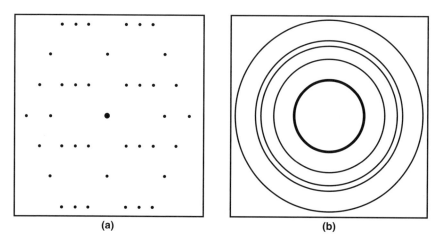

Figure 6.1 *X-ray diffraction patterns for (a) large single crystal, and (b) powder.*

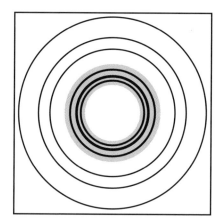

Figure 6.2 *X-ray diffraction pattern for some polymers displaying sharp rings superimposed on a diffuse background.*

polymer will crystallise. Polymers such as polyethylene $-(CH_2CH_2)_n-$, poly(ethylene oxide) $-(CH_2CH_2O)_n-$, poly(methylene oxide) $-(CH_2O)_n-$, etc. all have the regularity of structure required for the polymers to form an ordered structure. This also means that homopolymers have greater potential to crystallise than copolymers.

- *Geometrical regularity along the chain or stereoregularity.* Polymer chains should be able to adopt a conformation such that a simple translation of monomer units (or subunits) along the chain, or translation and rotation around the chain axis (to form a helix), yields identical units. We saw in Chapter 2 that polymerising vinyl monomers can result in three chain configurations, i.e. isotactic, syndiotactic and atactic. The first two types have geometrical regularity and therefore are likely to crystallise. The latter is unlikely to crystallise unless the size of the substituent group is small and close to the size of hydrogen atoms. An example of this is atactic poly(vinyl alcohol), in which the sizes of hydroxyl group and hydrogen atoms are similar. In this polymer the additional hydrogen-bonding interactions between the hydroxyl groups aid the creation of local order. Hydrogen bonding is very important in many polymer systems which exhibit crystalline characteristics, e.g. nylon and polyurethanes.

Polycondensation polymers, such as polyamides and polyesters, have geometric regularity by nature. Since they can have only one configuration, they are therefore semi-crystalline.

6.3 Polymer crystal structures

Using X-ray diffraction, crystallographers can measure the size of the unit cell for different semi-crystalline polymers without having to know how the crystals are formed. To understand the crystal structure of different polymers, one can imagine the following. Polymer chains have to be straightened and adopt the lowest energy; that is usually the *trans* conformation. This corresponds, for example, to a planar zigzag for polyethylene and to helices for isotactic polymers. Furthermore, the helical structure of isotactic polymers will change according to the size of substituents. Such regular chain structures can then aggregate side by side to form larger crystals. The pattern of repetition in all three dimensions will then define the unit cells of the polymers (Figure 6.3).

To describe the pattern of repetition, two sets of values are required. These are axes or distances (a, b, c) and angles (α, β, γ) as shown in Figure 6.4. Table 6.1 shows these parameters for some common polymers.

From the table, it can be seen that there can be situations where one polymer can form more than one type of crystal structure. This is called *polymorphism*. Polymers having different crystal structures will have different physical properties.

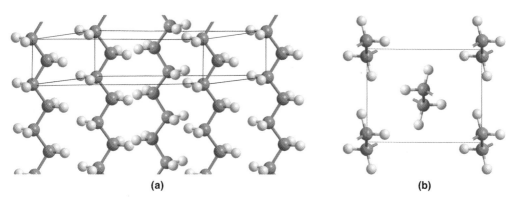

(a)   (b)

Figure 6.3 *Packing of chains (a) side by side, and (b) unit cell formation.*

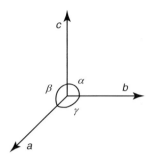

Figure 6.4 *Structural parameters (distances and angles) for describing the pattern of repetition or unit cell.*

Table 6.1 *Crystal structure and unit cell dimensions for some common semi-crystalline polymers*

Macromolecule	Crystal system Space group [1]Mol. helix	[2]Unit cell axes	Unit cell angles	No. units	ρ_c (g cm^{-3})
Polyethylene I — CH_2 —	Orthorhombic Pnam 1★2/1	7.418 4.946 2.546★	90° 90° 90°	4	0.997
Polyethylene II — CH_2 —	Monoclinic C2/m 1★2/1	8.09 2.53★ 4.79	90° 107.9° 90°	4	0.998
Polytetrafluoroethylene I — CF_2 —	Triclinic P1 1★13/6	5.59 5.59 16.88★	90° 90° 119.3°	13	2.347
Polytetrafluoroethylene II — CF_2 —	Trigonal P3$_1$ or P3$_2$ 1★15/7	5.66 5.66 19.50★	90° 90° 120°	15	2.302
Polypropylene (iso) — CH_2 — $CHCH_3$ —	Monoclinic P2$_1$/c 2★3/1	6.66 20.78 6.495★	90° 99.62° 90°	12	0.946
Polystyrene (iso) — CH_2 — CHC_6H_5 —	Trigonal R$\bar{3}$c 2★3/1	21.9 21.9 6.65★	90° 90° 120°	18	1.127
Polypropylene (syndio) — CH_2 — $CHCH_3$ —	Orthorhombic C222$_1$ 4★2/1	14.50 5.60 7.40★	90° 90° 90°	8	0.930
Poly(vinyl chloride) (syndio) — CH_2 — $CHCl$ —	Orthorhombic Pbcm 4★1/1	10.40 5.30 5.10★	90° 90° 90°	4	1.477

Table 6.1 *(Continued)*

Macromolecule	Crystal system Space group [1]Mol. helix	[2]Unit cell axes	Unit cell angles	No. units	ρ_c (g cm^{-3})
Poly(vinyl alcohol) (atac) —CH$_2$—CHOH—	Monoclinic P2/m 2*1/1	7.81 2.51★ 5.51	90° 91.7° 90°	2	1.350
Poly(vinyl fluoride) (atac) —CH$_2$—CHF—	Orthorhombic Cm2m 2*1/1	8.57 4.95 2.52★	90° 90° 90°	2	1.430
Poly(4-methyl-1-pentene) (iso) —CH$_2$—CH— | CH$_2$—CH(CH$_3$)$_2$	Tetragonal P$\bar{4}$ 2*7/2	20.3 20.3 13.8★	90° 90° 90°	28	0.822
Poly(vinylidene chloride) —CH$_2$—CCl$_2$—	Monoclinic P2$_1$ 4*1/1	6.73 4.68★ 12.54	90° 123.6° 90°	4	1.957
1,4-Polyisoprene (*cis*) —CH$_2$—CCH$_3$=CH—CH$_2$—	Orthorhombic Pbac 8*1/1	12.46 8.86 8.1★	90° 90° 90°	8	1.009
1,4-Polyisoprene (*trans*) —CH$_2$—CCH$_3$=CH—CH$_2$—	Orthorhombic P2$_1$2$_1$2$_1$ 4*1/1	7.83 11.87 4.75★	90° 90° 90°	4	1.025

[1]Notation for molecular helix has the form $A*u/t$, where A is the number of skeletal atoms in a repeat unit of the chain, u is the number of these units required for the crystallographic (chain direction) repeat distance and t is the number of turns of the helix in this crystallographic repeat.

[2]In unit cell axes indicates the chain axis.

6.4 Morphology of crystalline polymers

One of the very important questions in the formation of ordered structure in solid polymers is how such long molecules are able to form small crystallites (as deduced from X-ray scattering). The paradox is that the dimensions of the crystallites are much smaller than the lengths of the fully extended molecule.

The first attempt at an explanation of the structure of a crystalline polymer is due to Hermann and is called the *fringe micelle model*. According to this model, only part of the polymer molecule is incorporated into the crystalline phase, and the remainder forms a disordered region located around the edges of the crystal. The bits between these ordered regions have coiled conformations and hence form an amorphous region. The model proposes that a polymer meanders through a crystalline region, then into an amorphous region. The polymer chain

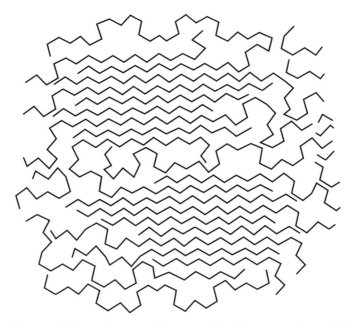

Figure 6.5 *Schematic drawing of the fringe micelle model.*

might come back to the same crystalline region or go into another crystalline region. The process repeats itself, and so we obtain many small crystalline regions connected by amorphous regions (see Figure 6.5).

This structure is consistent with X-ray powder patterns and other physical properties, thus was well received when it was first proposed.

However, the fringe micelle model is now known to be incorrect. Morphological studies of solid semi-crystalline polymers using electron microscopy reveal instead layer-like crystallites which are separated by disordered regions. This is called a *lamellar two-phase structure*. The thickness of these layer-like crystallites is in the 10 nm range and depends on the conditions used to grow the crystals. Long polymer chains form an essentially planar and thin structure by folding back and forth upon themselves. Such morphology is known as a *folded-chain lamella*. Figure 6.6 illustrates the structure.

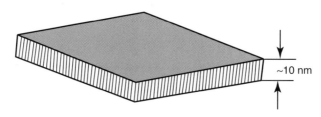

~10 nm

Figure 6.6 *Schematic drawing of the folded-chain lamellar model.*

On a macroscopic scale as seen under an optical microscope, semi-crystalline polymers also show another type of ordered structure. This comprises tiny spheres which are known as *spherulites*.

If we observe the crystallisation of a polymer melt under a polarised light microscope, we will see that the process begins with the formation of small nuclei somewhere in the melt. The nuclei then grow in size. This process is usually called *nucleation and growth*. The spherulites grow with a constant rate up to a point when they touch each other. The area of contact between two spherulites will be planar if the two were nucleated at the same time, and bent if the times were different. Finally the whole volume is covered by bound spherulites. The frequency with which nucleation occurs, which is referred to as the nucleation density, determines the final size of the spherulites. The sizes can vary over a large range, from a few micrometres to several millimetres.

Nucleation, in principle, occurs spontaneously when two or more polymer chains form a stable aggregate, but in most commercial samples is seeded by the presence of residual polymerisation catalysts or other foreign bodies in the melt. When the sample is ultra pure, nucleation is termed homogeneous, and when it is initiated by foreign bodies, it is termed heterogeneous. Homogeneous nucleation depends on two or more chains forming a stable aggregate, whereas heterogeneous nucleation involves adsorption of a polymer chain on the surface of the foreign body.

So polymer chains form both small layer-like crystallites or lamellae and larger spherulitic structures. The question now is how the two structures coexist at the same time in the same system. Since the lamellae are much smaller than the spherulites, they are clearly part of the spherulites. Figure 6.7 shows how the lamellae form the nucleus and grow into spherulites.

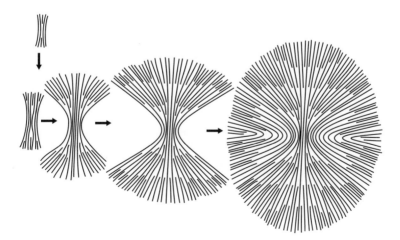

Figure 6.7 *Schematic drawing of the formation of nucleus from lamellae.*

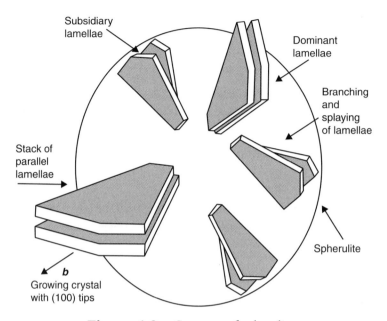

Figure 6.8 *Structure of spherulites.*

At an early stage of development, lamellae stack up and form a sheaf-like structure at the centre. This nucleus grows by curving out and this finally becomes a
stable spherical growth. To fill the space within a growing spherulite, branching
and splaying of lamellae occur. The lamellae orientate themselves in such a way
that their surface is perpendicular to the radius of the spherulites. Birefringence
measurement tells us that the polymer chains are also aligned perpendicular to the
radius of the spherulites. The final structure of spherulites is shown in Figure 6.8.

The extent to which crystallisation can occur is called the *degree of crystallisation*
or the crystallinity. There are two slightly different ways for defining crystallinity
and these reflect the method used in its determination. It is usually assumed that
the system is a two-phase structure, i.e. crystalline and non-crystalline. If we measure the density of a sample, we can write that the total ensemble is described by:

$$V\rho = V_a\rho_a + V_c\rho_c$$

where V_a and V_c are the volumes occupied by the amorphous and the crystalline
phases, respectively. V is the total volume. ρ_a, ρ_c and ρ are the respective densities.

The crystallinity can be defined as the volume fraction of the crystalline phase.
This is denoted by ϕ_c and we write:

$$\phi_c = \frac{V_c}{V}$$

So ϕ_c can be rewritten as:

$$\phi_c = \frac{\rho - \rho_a}{\rho_c - \rho_a}$$

To evaluate the crystallinity from this equation, we need to know the densities of the crystalline and amorphous phases. The value of ρ_c can be derived from the lattice constants determined by X-ray diffraction. The value for ρ_a is less easy to define as it reflects a completely disordered phase. Usually ρ_a is obtained by an extrapolation of the melt density assuming a constant coefficient of thermal expansion.

Another more convenient method of determining the crystallinity is to measure the heat of fusion, ΔH_f, of a sample. This leads us to use the weight fraction of the crystalline phase rather than the volume fraction as above. The crystallinity by weight, ϕ_c', can be determined from the heat of fusion, ΔH_f, by:

$$\phi_c' = \frac{\Delta H_f}{\Delta H_f(\phi_c' = 1)}$$

To use the equation, the value of the heat of fusion of a fully crystalline sample is required. Since such a sample cannot be prepared, an extrapolated value has to be used.

6.5 Mechanisms of crystallisation

Now how does crystallisation occur? There can be very different starting conditions for crystallisation. The usual, and therefore most important, is during solidification of an isotropic melt. However, crystallisation can also occur in the solid state when glassy polymers are heated to just above their glass transition temperatures. The morphologies obtained by the two methods might well be different.

In the melt state, polymer chains exist in a coiled structure and are mutually interpenetrating. The shape of the polymer coil will be continuously changing. When the melt is cooled, the segments which have the lowest energy helical conformation will increase and the coil will become less entangled. Nucleation, which is the initial part of the process which transforms the melt to a solid, involves the linear sections aggregating and then packing regularly to form small crystals. This initial process in which the isotropic melt is transformed to a sizable nucleus is called *primary nucleation*. Once this sizable nucleus has been formed, it grows spontaneously. From these steps, the crystallisation isotherms have sigmoidal shapes as shown in Figure 6.9.

The first part of the curve is an induction period in which time is required for the formation of nuclei. The steepest part is related to the growth of the spherulites. The final part is when the spherulites begin to touch each other and the rate slows down. When the crystallisation temperature is increased, the shape

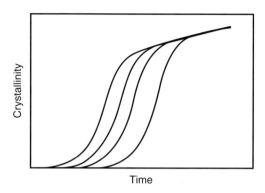

Figure 6.9   *Crystallisation isotherms at different temperatures.*

of the curve remains the same but shifts to a longer induction time. The last part of the curve is known as *secondary crystallisation*, and this continues over time.

All the above processes require substantial times, so usually crystallisation is not completed before further chain reorientation becomes impossible. So entanglement and disordered regions remain in the system. This means that polymers are never perfectly crystalline. The internal energy is therefore not at the lowest state and it can be said that crystallisation is governed by kinetic criteria rather than by equilibrium thermodynamics.

The coiled form of the polymer has a large number of the higher energy conformations because thermodynamics dictates that these are populated at high temperature. As the melt is cooled, the number of these conformations will be decreased and the proportion of the chain in the lower energy linear conformation will increase. The net effect is that the polymer coil will expand and become less coiled. In fact, the size of the lamellae will reflect the temperature at which crystal growth occurs. The lower the temperature at which crystallisation occurs, the longer the sections which are linear and the thicker the lamellae.

If the melt is not disturbed during solidification, crystallisation normally results in the spherulitic morphology described above. In reality, and often in many special cases, the melt is disturbed, both unintentionally and intentionally, resulting in a special morphology. Also, attempts have been made to "direct" the crystallisation to specific structures with required properties. When the melt is either stretched or sheared, the chain molecules will be deformed and stretched. This brings the chains closer to the extended state seen in crystals. As a consequence, the induction time for nucleation is shorter. Further, the nucleus changes in shape to become long, needle-like entities aligned in the direction of stretching or shear. The rest of the molecules can then deposit in the usual folded-chain fashion on these long nuclei. The resulting morphology is called *row crystallisation* and *shish-kebab*.

So the morphology found in semi-crystalline polymer solids is a reflection of the conditions used in the production of the solid: these are the temperature of the melt from which cooling is initiated, the rate of cooling, and whether or not the liquid experiences shear or compressive forces during solidification.

Crystallisation can also occur in the solid state. This is usually associated with semi-crystalline materials that crystallise rather slowly and therefore can be quenched into a glassy solid. A well known example of this type of polymer is poly(ethylene terephthalate). Upon heating to just above its glass transition temperature, crystallisation starts. This is known as *cold crystallisation*.

We reiterate that crystallisation is a kinetic process. If given enough time, a polymer with an appropriate structure (see section 6.2) can crystallise. However, whether or not it will crystallise upon cooling is a different matter. The nucleation and growth, and hence the resulting crystallinity, are determined by how easily the chain can adopt the necessary conformation. As was discussed earlier, this is determined by the energy barrier for rotation around a particular bond. If the energy barrier is high then the induction period for the formation of nuclei will be long and growth will be slow. The crystallisation might well stop if the temperature drops faster than the crystals are growing. As a result, the crystallinity will be low. In such systems it is possible to quench the melt fast enough to obtain glassy amorphous polymers.

Crystallisation has also been described as spinodal decomposition. This originally explained phase separation of a solution of two or more components into distinct regions with different chemical compositions and physical properties. The polymer melt can be thought of as a solution of crystallisable and non-crystallisable components. The latter can be the same polymer as the former but contains some defects such as branches or chemical irregularities. The crystallisation proceeds by diffusion together of the same species to form regions rich in one component. This process occurs uniformly throughout the system. The concentration gradient driving the diffusion is very low at the beginning of the crystallisation, rising to pure component as time proceeds. This separation process is determined solely by diffusion, not by the thermodynamics of fluctuation as is the case in nucleation and growth.

Regardless of crystallisation mechanism, if a melt is cooled slowly, the molar mass effects can influence the nature of the morphology which is created. High molar mass material has a higher meting point and will tend to solidify first from the melt. Lower molar mass material will hence be phase segregated and will separate into smaller crystals. A detailed study of the melting of commercial polymer materials often indicates that rather than one melting point there may be a distribution of melting events reflecting the molar mass distribution of the material. The extent to which this phase segregation occurs depends on the time involved in the cooling of the melt and again reflects the kinetic nature of the processes involved.

6.6 Temperature and growth rate

Growth rates vary strongly with temperature. Generally, there is a maximum growth rate at an optimum temperature (Figure 6.10). The growth rate decreases rapidly on both sides of this, towards the equilibrium melting point at high temperatures and towards the glass transition at low temperatures. On the low temperature side, the rate drops because of the slowing down of the segmental mobility in the melt. On the high temperature side, the segmental motion becomes greater and the condensation to stable nuclei decreases as temperature increases.

To obtain a significant number of nuclei having the critical dimensions required for stability and further growth, the melt has to be supercooled by about 5 °C to 20 °C. A nucleating agent can be added to induce crystallisation at the higher temperatures. This is known as *heterogeneous nucleation*. At a given tempera-

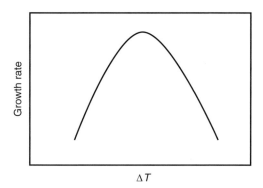

Figure 6.10 *Schematic of growth rate vs. temperature difference from its melting temperature.*

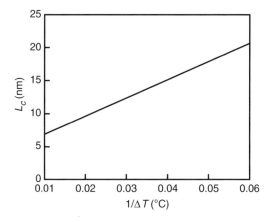

Figure 6.11 *Effect of crystallisation temperature on lamellar thickness.*

ture, a nucleating agent increases the crystallisation rate by shortening the induction or nucleation period but does not affect the spherulitic growth rate.

The crystallisation temperature affects not only the crystallisation rate, but also the lamellar thickness. Increasing the crystallisation temperature will result in thicker lamellae, as shown in Figure 6.11.

6.7 Melting of crystalline polymers

For a perfectly crystalline substance, melting is characterised by a marked volume change and a well defined melting temperature. The change is usually referred to as a first-order transition. Since polymers are not perfectly crystalline, but contain disordered regions and crystallites of varying size, melting takes place over a range of temperatures. The range of melting temperature of a polymer is indicative of the size and perfection of the crystallites in the sample. A good example of this is the melting of polyethylene, which can exist in both linear and branched structures. Linear or high density polyethylene (HDPE) has a higher crystallinity, sharper melting range, and higher melting temperature than branched or low density polyethylene (LDPE). Imperfections on the chain reduce the crystallinity, broaden the melting range, and reduce the melting temperature.

In addition large crystallites have a higher melting temperature than small ones. The relationship between the thickness, l, of a crystal and its melting temperature, T_m, is known as the Thomson–Gibbs equation:

$$T_m = T_m^0 \left[1 - \frac{2\sigma}{l\rho_c \Delta H} \right]$$

where T_m^0 is the equilibrium melting temperature, σ is the surface free energy per unit area and ΔH is the increase in enthalpy per unit mass on melting for an infinitely thick crystal. It can be seen in this equation that the highest possible melting temperature is the equilibrium melting temperature. It is defined as the melting point of a crystal which is so large that the size effects (surface) are negligible, and which is in equilibrium with the normal polymer liquid.

The crystallisation temperature determines the thickness of the lamellae, and then the melting temperature depends on the thickness of the lamellae. So by programming the crystallisation temperature, one can obtain materials with quite peculiar melting patterns, such as multiple melting peaks. Phenomena like these, of course, are not observed in small molecules.

Some polymers can undergo crystallisation in what is apparently the solid state (for example cold crystallisation in poly(ethylene terephthalate)). Molecular motion of specific parts of the polymer chains occurs and molecules can then adopt the right conformation to form a crystal. In a similar manner, a polymer that has already crystallised to a certain lamellar thickness (determined by crystallisation temperature) can be heat treated so that the lamellae become thicker. The heat treatment, or annealing, process involves heating the polymer to high enough temperatures to allow molecular motion, but low enough not to melt and destroy the crystals. The process is known as *lamellar thickening*.

Further reading

Mandelkern L. *Crystallization of Polymers*, Cambridge University Press, Cambridge, 2002.

Vaughan A.S. and Bassett D.C. *Crystallization and Morphology*, in G. Allan and J.C. Bevington (Eds.), *Comprehensive Polymer Science, Volume 2 Polymer Properties*, Pergamon Press, Oxford, 1989.

7

The rubber state

The rubber state is unique to linear and lightly cross-linked polymeric materials. It is characterised by the ability to undergo very large reversible deformations, to recover its original shape, to absorb energy and to damp vibrations.

7.1   Large deformations

The ability of a sample to undergo large deformations is a reflection of the fact that the molecules within the sample can themselves experience huge changes in shape. This is made possible by the summation of internal rotations in the many units making up the polymer chain. Thus a molecular chain for which the unstressed state, called the *relaxed* state, is to be coiled up into the contour of a ball, under the influence of a deforming stress is pulled out into the contour of an ellipse. For large chains, extensions of several hundred per cent are possible. The stresses required for this are less than those required to deform the glassy state, and so the rubber is characterised by a much lower modulus than the glass. Most rubbers will have a modulus of ~10^6 Pa. The maximum possible extension occurs when sections of the chain are pulled completely into the planar zigzag determined by the covalent bond angles (Figure 7.1).

In practice, the individual polymer coils will rarely be fully extended. In most thermoplastic samples, the polymer chains will be entangled and the degree to which full extension can be achieved is influenced by the presence of these physical entanglements. In cross-linked rubbers, the chemical cross-links will determine the extent to which the chains between the cross-links can be extended. However, the concept of the transformation of a coiled chain to one that is fully extended is a good starting point for consideration of the changes in shape which occur in elastomeric materials.

7.2   The elastic restoring force

Examining the two molecular conformations pictured above, it is possible to identify the source of the restoring force.

First, the coiled shape can adopt many different conformations, all giving the circular contour. However, when the chain section is pulled completely out into the zigzag constrained by the covalent bond angles, there is only one possible

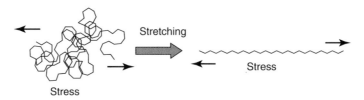

Figure 7.1 *Large molecular deformations.*

conformation. Thus the change is marked by a reduction in the number of possible conformations that the chain can adopt.

Now consider Boltzmann's definition of the disorder of a system, using the thermodynamic measure of *entropy*. Boltzmann's equation relates entropy to the number of possible arrangements of a system by:

$$S = k \ln(\textit{number of arrangements})$$

where S is the entropy, k is Boltzmann's constant and we shall use the number of permitted conformations as the number of arrangements. Then the entropy change on stretching the molecules of a sample is:

$$\Delta S = k \ln(\text{number of conformations})_s - k \ln(\text{number of conformations})_u$$

where subscripts s and u refer to stretched and unstretched states respectively. So, denoting the number of conformations by the Greek letter Ω:

$$\Delta S = k \ln(\Omega_s / \Omega_u)$$

This equation shows us two things. Firstly, the process of stretching a polymeric material gives rise to a negative entropy change, since Ω_s is less than Ω_u. Secondly, in principle we can count the number of possible arrangements that the chain can adopt and so calculate the entropy of stretching.

The consequence of the negative entropy change can be found in the familiar rules of thermodynamics. Nature prefers disordered systems of high entropy to ordered systems of low entropy. So the direction of spontaneous change for a system is towards the state of higher entropy. To be more precise, the direction of spontaneous change is towards the state of lower free energy:

$$\Delta G = \Delta H - T \Delta S$$

where G is the free energy, H is the enthalpy and T is the temperature. However, for the change in shape of a rubber molecule, the energy/enthalpy change is usually considerably less than the entropy component, and we can say for the free energy of stretching:

$$\Delta G_s = - T\Delta S_s$$

Since the entropy change is negative, the free energy change on stretching is positive. Thus the direction of spontaneous change is from stretched to unstretched. This is the thermodynamic origin of the restoring force. It is important to note that it is an *entropic force*, whereas the forces restoring deformation in other familiar solids are the result of energy/enthalpy increases on deformation. Two significant consequences of this arise. Firstly, since the entropy decreases on stretching, the restoring force increases with the strain. Secondly, since the entropy change always occurs in the free energy equation multiplied by temperature (as $T\Delta S$), the restoring force in a stretched rubber, at constant strain, increases with increasing temperature. This is in direct contrast to other familiar solids, when the energy-derived restoring force is either independent of, or decreases slightly with, temperature.

We can illustrate these effects with the following example. Consider a weight suspended from a hook on a rubber band. The mass of the weight stretches the rubber. If we now use a hot-air blower to heat the stretched rubber, what happens?

The stretched length, l_s, is a balance between the downward force exerted by the weight and the restoring force of the rubber. This balance occurs because the restoring force rises with the amount of deformation. In fact, the relationship between extent of deformation and restoring force approximately follows Hooke's law, as will be shown below. If we now heat the stretched rubber, the restoring force rises and becomes greater than the mass of the weight, and so the stretched length decreases, decreasing the new restoring force until once again it exactly balances the mass of the weight. So the weight *rises* as indicated by the upward arrow in Figure 7.2.

In principle, for any given molecular structure, we can count the number of possible conformations and so use Boltzmann's equation to calculate the free energy of stretching and so its differential with respect to distance, which is the restoring force. However, the numbers of arrangements are very large and it is not practical to count the actual numbers. Mooney and Rivlin have formulated an equation which captures the essential features of the processes which are occurring. The Mooney–Rivlin treatment essentially follows these statistical principles, and arrives at an equation for the restoring force of great practical usefulness:

$$-F = \frac{\rho RT}{M_c}\left\{\frac{l_s}{l_u} - \left(\frac{l_u}{l_s}\right)^2\right\}$$

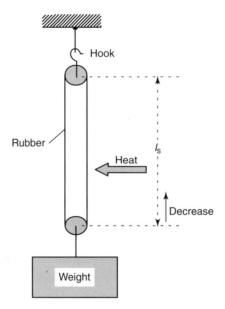

Figure 7.2 *Stretched rubber contracts on heating.*

In this equation, the negative value of F simply indicates that the force acts in the opposite direction to that of the strain, ρ is the density accommodating the number of deformed chains per unit volume of rubber, l_u and l_s are the unstretched and stretched lengths respectively, R is the gas constant, deriving from Boltzmann's constant, k, as we move from molecular to macroscopic quantities, M_c is the molecular weight of chain between cross-links, the units by which the macroscopic applied stress is applied to the deforming parts of the molecule.

Consider the Mooney–Rivlin equation in a little more detail. First, at significant extensions, $(l_u/l_s)^2$ can be ignored in comparison with l_s/l_u, so that the force becomes proportional to the extension and the sample follows Hooke's law. The curvature at low strains, due to the $(l_u/l_s)^2$ term, accommodates the several assumptions in the derivation of the equation, including the neglect of the energy of conformation change. It also, of course, ensures that the force is zero at zero strain, and not $\rho RT/M_c$. In practice, though, the Mooney–Rivlin equation overestimates the stress at this low strain part of the curve. In fact, sometimes the stress even decreases with strain rather than increasing! To explain this we assume that not only are the chains tied to each other by permanent chemical bond cross-links, but they also intermingle, forming a number of knots or entanglements. These are not permanent, but can slip or undo under the influence of the stretching force. An even better fit of theory to experiment is obtained if the statistical counting procedure examines not one chain section between junction points, but a matrix of several such sections. However, such refinements of statistical counting and thermodynamic theory are beyond the scope of this volume.

At very high strains, the molecules are starting to become extended, and can then pack neatly together and so crystallise. The modulus of the crystalline material rises sharply, and this area of the stress–strain curve is called *stress crystallisation*. Stress crystallisation can be observed as whitening of the polymer when materials are stretched. The whitening is the visible evidence of the chains becoming well ordered and light being scattered by these ordered regions (Figure 7.3).

Continuing this examination of the terms in the Mooney–Rivlin equation, we note the inclusion of temperature, as we have discussed above, carrying through from the $T\Delta S$ origin of the force. So here there is a formulation of the characteristic of contraction on heating that we noted earlier.

Finally, there is the inverse dependence on M_c, the molecular weight of the chain constrained between cross-links. This is also surprising at first sight. Suppose that we "take hold" of and "pull" the chain by the cross-links, and the section between links is the length of chain being deformed. Then, going back to Boltzmann's equation, we remember that the change in entropy depends on the ratio of the number of conformations, not the simple difference. Then the relative decrease in the number of conformations is greater for a short chain than for a long one. To see this, consider ten conformations being reduced to five, a relative decrease of 50%. On the other hand, 1000 conformations being reduced by the same amount to 995 is a relative decrease of only 0.5%.

The Mooney–Rivlin equation is of great practical use in rubber technology, especially for relating the restoring force to factors such as the degree of cross-linking

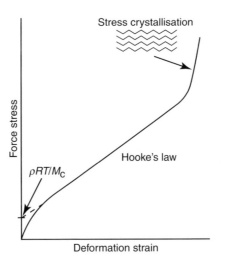

Figure 7.3 *Stress crystallisation.*

(vulcanisation). Of course other things, such as the elastic modulus, also change with the degree of cross-linking, so this describes only one aspect of rubber behaviour. In general, the higher the degree of cross-linking, the higher the value of the modulus for the rubber.

7.3 Energy absorption

The third important property of rubber is its capability to absorb energy from a stress–strain cycle, a phenomenon known as energy loss or energy damping. To introduce this, again ask the question: "In order to deform, what do the molecules need?" Answer: "The molecules need energy."

So far we have considered that this energy is supplied (or not supplied) by thermal energy of value kT per rotating unit. But suppose that this energy is almost, but not quite, enough to permit the rotating unit to cross the rotational energy barrier. Or that the rate of crossing the barrier is just below that necessary to observe deformation in the time of observation. Then externally supplied energy in the form of a mechanical stress can combine with the thermal energy to "push" the chain over the barrier that restricts its rotation.

Required energy = thermal energy + applied energy

This applied energy is obtained in the form of work done on the system, on the molecules.

Required energy $= kT + W$

where W is the work done, energy lost by conversion to heat. We can visualise this as a molecular push (Figure 7.4).

Work done is force times distance:

$$W = F \times d$$

In the glass there is no movement of the polymer backbone and d is zero, so W is zero. In the rubber, thermal energy is adequate for facile backbone rotation and

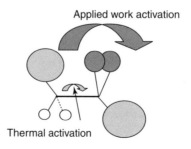

Figure 7.4 *Work plus thermal energy causes rotation.*

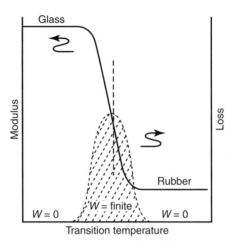

Figure 7.5 *Work done rises, then falls, with temperature.*

no force is necessary to promote movement so that F is zero and so again W is zero. In the region of the transition, additional energy helps the chain to rotate, so that F is finite, and there is movement when the stress is applied, so d is finite, and thus W is finite.

These three limiting conditions can be inserted onto our familiar modulus–temperature diagram, and joined up with a smooth curve, the exact shape of which will be derived later (Figure 7.5).

This bell-shaped curve that has been inserted without, as yet, any curve formula, is one that will become familiar for many different aspects of time dependent properties. Later it will be shown that both the sigmoid (drawn here for the modulus), and the bell-curve (here the loss), occur together as the coupled real and imaginary components of the mathematically complex functions describing periodic (alternating) phenomena.

It is important to note that, although the loss is a maximum at the point of inflection in the modulus curve, the tails extend considerably into both the glass and the rubber regions. It is this latter that gives rubbers their energy absorbing, vibration damping properties.

In this simple introductory treatment it is also possible to make a qualitative estimate of the magnitude of the loss. Since the work is done helping the rotating unit cross an energy barrier, the higher the barrier the higher will be the work required. So, as a rule of thumb, rubbers with large energy barriers will require more work and absorb more energy than chains with low energy barriers.

Now examine the consequences of this with three rubbers of technological significance: polybutadiene, natural rubber and butyl rubber.

Polybutadiene (PB)

$$CH=CH$$
$$—CH_2 \qquad CH_2—\ \ \text{unhindered}$$

cis-1,4-Polyisoprene (NR)

$$CH_3$$
$$C=CH$$
$$—CH_2 \qquad CH_2—\ \ \text{slightly hindered}$$

Poly(isoprene-co-isobutene)

$$CH_3$$
$$C=CH \qquad CH_3$$
$$—CH_2 \qquad CH_2—C—\ \ \text{very hindered}$$
$$CH_3$$

We look first at the energy–angle diagrams (Figure 7.6).

In this diagram, the energy scale is enlarged, so that the barriers involved are much less than those portrayed earlier for polystyrene and poly(vinylcarbazole). To understand the way in which energy is absorbed by the rubber, we must inspect the variation of the modulus and loss as a function of temperature (Figure 7.7).

Looking at the three dashed curves that describe the temperature variation of the loss for the three rubbers, polybutadiene has the lowest loss, with the transition temperature furthest below room temperature. Butyl rubber has the highest loss, with the transition temperature closest to room temperature. Natural rubber is between the two. The loss peak therefore moves on the temperature axis and the rubbers' ability to absorb energy will vary with the material. So at room temperature the loss rises in the sequence PB < NR < BR.

A low loss rubber, like polybutadiene, reacts to deformation by recovering rapidly and returning most of the energy applied by the original deforming stress.

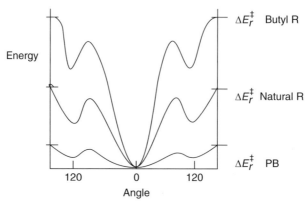

Figure 7.6 *Energy profiles for three rubbers.*

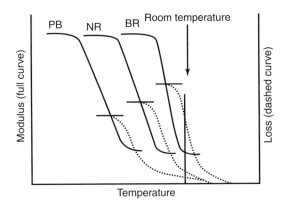

Figure 7.7 *Modulus and loss of three rubbers near room temperature.*

This property is called *resilience*. As a consequence, it does not heat significantly when stretched at room temperature. Objects with high resilience fly further when hit (as in golf balls) or bounce higher when dropped. A high loss rubber, like butyl rubber, absorbs much of the energy when deformed at room temperature. So objects constructed of such a rubber are used for vibration and sound damping, or where resilience and bouncing are undesirable. Thus by changing the chemical structure of the elastomer it is possible to vary its characteristics so as to shift the glass transition temperature, and hence the loss peak, along the temperature axis.

7.4 Tyre technology

The energy absorbing properties of a rubber are of tremendous importance in vehicle tyre technology. The different parts of a tyre are constructed of rubbers of different loss absorption.

Consider the cross-section of an unloaded tyre (Figure 7.8). The tyre is positioned on the wheel rim by the metal *beading*, the *carcase or body* of the tyre is

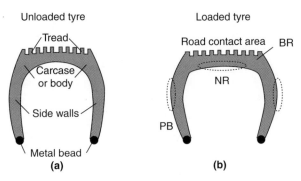

Figure 7.8 *Tyre cross-sections: (a) unloaded; and (b) loaded.*

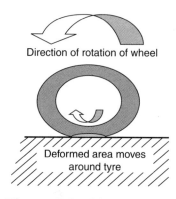

Direction of rotation of wheel

Deformed area moves
around tyre

Figure 7.9   *Tyre in motion.*

supported by the *sidewalls*, and the section in contact with the road is the grooved *tread*.

When the tyre is loaded by the weight of the vehicle, the sidewalls are deformed in the region immediately above the road contact area. As the wheel rotates, the contact area moves round the tyre, and so too does the area of deformation of the sidewall (Figure 7.9).

Consequently, as the tyre turns round, each area of the sidewall is alternatively deformed and then relaxed, a movement that creates an alternating stress/strain in the rubber. At each cycle, energy is absorbed, and the tyre temperature rises until the rate of heating by this energy loss is balanced by the rate of cooling due to radiative loss or conduction. If the deformation is excessive (due to overloading of the vehicle), and if the frequency of deformation is high (due to speeding), the rate of heating exceeds that of cooling and the temperature rises to the point where parts of the rubber weaken, or even melt, the tyre sheds its tread, explodes and the vehicle crashes (Figure 7.10).

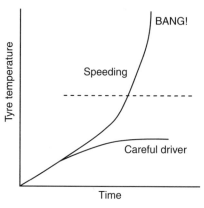

Figure 7.10   *Tyre heating.*

In order to minimise the heating, the tyre should be correctly inflated, the vehicle not overloaded, and the vehicle driven at a speed not above the maximum recommended by the tyre manufacturer.

For this reason, the sidewalls of a tyre are constructed of rubber with the least energy loss. When cost is not a factor, this means the use of a special low loss rubber such as polybutadiene. However, it is important to recognise that cost is often a factor, and many less expensive tyres have sidewalls constructed of cheaper natural rubber, with consequent higher heat generation and more severe restrictions on vehicle speed.

However, low loss rubbers like polybutadiene are extremely resilient, which means that articles constructed of them (like rubber balls) tend to bounce very readily. Obviously, we do not want a vehicle that bounces its way along the road, so for the contact tread area we need a rubber that absorbs shocks without bouncing, and so adheres to the road. For this purpose we need a high loss rubber, and again, where cost is not a factor, a special rubber, like butyl rubber, is used. Such rubbers give better *road adhesion*, particularly in the wet. Once again, it is necessary to note that cheaper tyres, with the tread constructed of natural rubber, are more prone to skid on wet roads than are those constructed with a tread of special high loss rubber.

Finally, for cost saving, the carcase, or body, of the tyre, which is neither deformed nor in contact with the road, can be constructed of the cheapest effective rubber, which is natural rubber.

So, a top quality tyre, as required for high speed luxury vehicles, is constructed of three different kinds of rubber in the sidewall (often polybutadiene blends), carcase (natural rubber) and tread (often butyl rubber blends). For a less expensive tyre, constructed totally of natural rubber, a similar (though lesser) effect can be achieved by using more flexible, lightly vulcanised, rubber in the sidewall and higher modulus, more extensively vulcanised, and so more tightly cross-linked, rubber in the tread.

In practice, a tyre will have an internal canvas and metal fibre structure which helps spread the load. The design of high performance tyres is not as simple as it might at first sight appear.

7.5 Cross-linked rubber

In use, most rubbers are lightly cross-linked so that the molecules cannot translate relative to each other. This prevents first creep, then, at higher temperatures, flow. Since flow is necessary in the processing of articles into their final shape, the cross-linking reaction is usually carried out using the elevated temperatures of the shaping process such as extrusion or moulding.

For natural rubber, by far the most important industrial cross-linking process is vulcanisation. The rubber is processed mixed with sulfur and a catalyst (as well as other additives and fillers like carbon black). Then short polysulfide chains add to the double bonds of the polyisoprene chain.

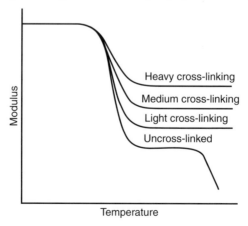

When a rubber is very lightly cross-linked, the modulus is barely affected, but the rubber region extends to higher temperatures and the creep/flow region is eliminated. However, as the cross-linking density is increased, the network becomes tighter and tighter, and shape change in the chains between the cross-links becomes more and more difficult. Then the modulus, the restoring force and the loss all rise (Figure 7.11).

An interesting form of cross-linking is found in what are called *thermoplastic rubbers*. These are block copolymers of a flexible rubber chain with a less flexible chain that is a glass at room temperature. Usually there are two blocks of hard polymer separated by a block of flexible polymer. These have already been described in Chapter 2. A typical example contains first a chain of polystyrene units connected to a chain of polyisoprene units, in turn connected to a second chain of polystyrene. Now polystyrene and polyisoprene are not miscible, so the material undergoes a phase separation into polystyrene and polyisoprene phases.

Figure 7.11 *Modulus of cross-linked rubber.*

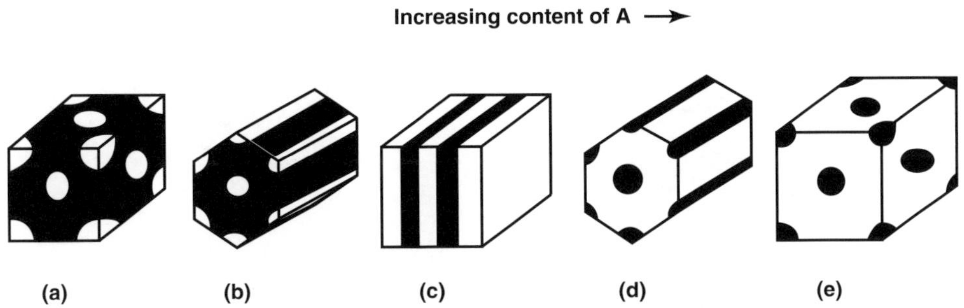

Increasing content of A ⟶

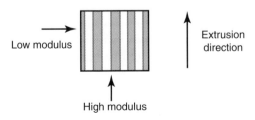

(a)   (b)   (c)   (d)   (e)

Figure 7.12 *Effect of composition on block copolymer morphology: (a) spheres of A in matrix of B; (b) cylinders of A in matrix of B; (c) alternating A and B lamellae; (d) cylinders of B in matrix of A; and (e) spheres of B in matrix of A.*

If the relative chain lengths are chosen correctly, this can result in separated zones of glassy polystyrene in a continuous matrix of polyisoprene. The polystyrene phase can be spheres, cylinders or laminae (sheets), again depending on the relative volume fractions of the two types of chain (Figure 7.12).

Since the ends of the polyisoprene chains are covalently bonded into the polystyrene glass, they cannot move away, and the glassy polystyrene phase effectively "cross-links" the rubber polyisoprene chains. However, at temperatures above the glass transition of polystyrene, the chains can be pulled out of the polystyrene domains, and so the article can be reshaped, becoming "cross-linked" once again on cooling.

If the glassy phase consists of cylinders or sheets, these become oriented in the sample during the processing and formation processes. As a result, the mechanical properties of the sample are anisotropic. For example, for glassy cylinders in a rubber matrix, the modulus in the cylinder direction is the high modulus of glass, whereas in the two perpendicular directions the moduli are the low values of a rubber. Similarly, for glassy sheets, the modulus is high in the two sheet dimensions, but low in the perpendicular direction. This anisotropy is used in high technology applications where different mechanical properties are required in different directions (Figure 7.13).

Low modulus ⟶

Extrusion direction ↑

High modulus ↑

Figure 7.13 *Anisotropic modulus of extruded thermoplastic rubber.*

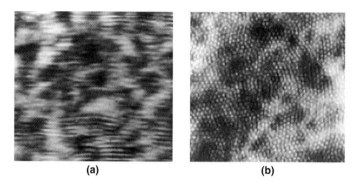

(a)   **(b)**

Figure 7.14 *Electron micrographs of osmium tetroxide stained styrene/butadiene/ styrene triblock copolymer obtained in (a) the extrusion direction; (b) transverse to the extrusion direction.*

Electron micrographs of stained styrene/butadiene/styrene (SBS) triblock copolymer are shown in Figure 7.14.

Osmium tetroxide reacts with the butadiene and stains the material dark when imaged using an electron beam. The styrene phase (shown as white) is seen in the transverse direction as circular domains and when viewed in the longitudinal direction appears as stripes. The styrene is thus present as cylinders and gives the material a high modulus in the one direction, whereas in the transverse direction the stress is taken by the black rubbery butadiene phase.

7.6 Two other significant elastomeric materials

A number of polymeric systems exhibit elastomeric properties. In each case, the molecular motion characteristic of rubbers is sensitive to the molecular and supramolecular structure. Two interesting, and commercially important, examples worthy of further discussion are polyurethanes and polysiloxanes, or silicone rubbers.

7.6.1 Polyurethanes

The term polyurethane (PU) covers a number of polymer systems with one common element of chemistry, they contain an −NH−CO−O− linkage. Polyurethanes are used in a variety of applications where the need to form and mould the product into a desired shape makes the use of natural rubber based materials difficult. Such products include soles of shoes such as trainers, conveyor belts, foamed car seats and furniture, wiper blades in printers, etc. By changing the formulation it is possible to change a PU from being a hard material to being a very soft material.

Thermoplastic PU rubbers, as the name implies, are polymers which are not cross-linked yet exhibit many of the characteristics of cross-linked rubbers. To form these polymers, a diisocyanate is reacted with a polyether, or a polyester, diol to produce the polyurethane. Two common types of isocyanate used are 4,4′-methylenediphenyl diisocyanate (MDI) and mixtures of 2,2′-methylenediphcnyl diisocyanate and 2,4–toluene diisocyanate (TDI).

Methylenediphenyl diisocyanate (MDI)   Toluene diisocyanate (TDI)

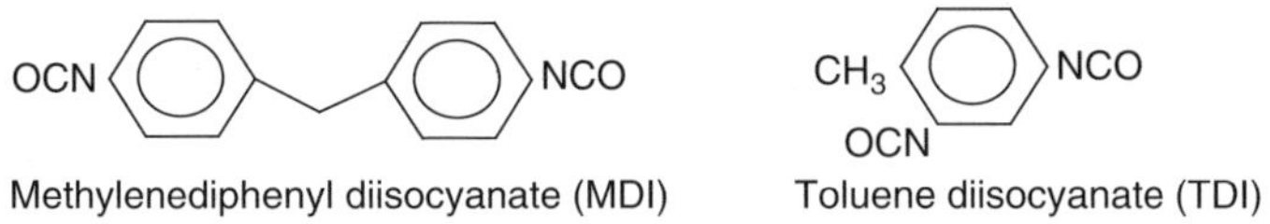

TDI is usually used only in foam products as it has a higher level of toxicity than MDI and hence is less easily handled. By changing the nature of the poly-ether or polyester sections of the chain, the susceptibility of the material to oil and water can be altered. Polyethers can absorb water, whereas polyesters with large aliphatic chains between the ester groups are less susceptible to moisture uptake. The typical thermoplastic PU rubber is created by the chain extension reaction of a preformed isocyanate capped monomer, usually based on MDI. The chain extension reaction quickly produces a phase separated product. The urethane segments aggregate together and are effectively hydrogen bonded with neighbouring molecules to form a stable structure.

The polymer chain then contains flexible and less flexible blocks (Figure 7.15).

The urethane "hard" phase has a melting point of approximately 150°C. The linking groups are selected to have glass transition temperatures which are below room temperature. The polyether/polyester rich phase is often referred to as the "soft" phase, reflecting the fact that at room temperature it has mobility, giving it

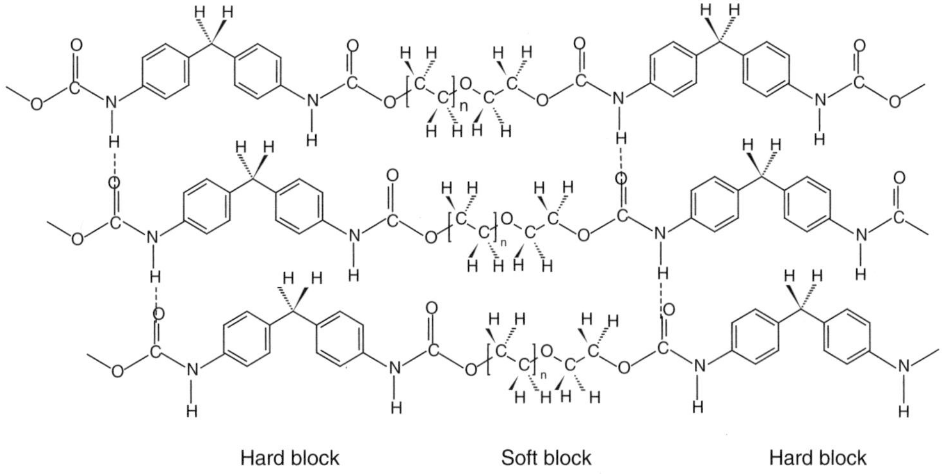

Hard block   Soft block   Hard block

Figure 7.15   *Schematic of a polyurethane elastomer.*

elastomeric characteristics. So in general terms, the chain molecular motion resembles that in the three-block copolymers described in section 7.5 above, with the hard blocks behaving as the thermoplastic cross-links. However, in contrast to the styrene/butadiene/styrene three-block copolymers, where the moulding temperature is defined by the T_g of the polystyrene phase, in PU it is the melt temperature of the hard phase which is the defining characteristic.

Interestingly, the selection of the molar mass of this linking block is very important and can dramatically influence the properties of the material. If the molar mass is very low then the distance between the urethane blocks is short and the material is hard, and the soft block is constrained with a T_g value which is often close to or can be above ambient temperature. As the molar mass of the linking unit is increased, so the chains are less restricted and T_g is lowered. For a simple poly(ethylene oxide) (PEO) of molar mass 2500 this can drop to around $-20°C$. This is ideal for shoes, which, if the T_g were too high, would become inflexible in cold weather. However, if the molar mass is increased to 4000 or above, the PEO chains in the "soft" phase are able to interact and may crystallise. Then once more the molecular motion of the chains is constrained and the T_g is raised. Typical modulus–temperature plots for a PU are shown in Figure 7.16.

Because the temperature has to be raised above 150°C before the molecules forming the "hard" block can move, the material is dimensionally stable up to these temperatures, significantly higher than in the hydrocarbon thermoplastic elastomers. However, because of the chains which are above their T_g in the "soft" phase, excellent elastomeric properties can still be achieved. A well designed PU can be extended up to 600% before it will break. This is comparable to the extension that can be achieved with natural rubber.

Another striking feature of the PU rubbers is the very extensive temperature range over which an almost constant modulus is maintained. In the above example, the modulus drops to a value of approximately 10^7 Pa at the lower glass

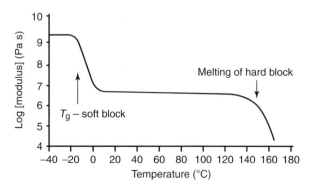

Figure 7.16 *Modulus–temperature plot for a typical polyurethane.*

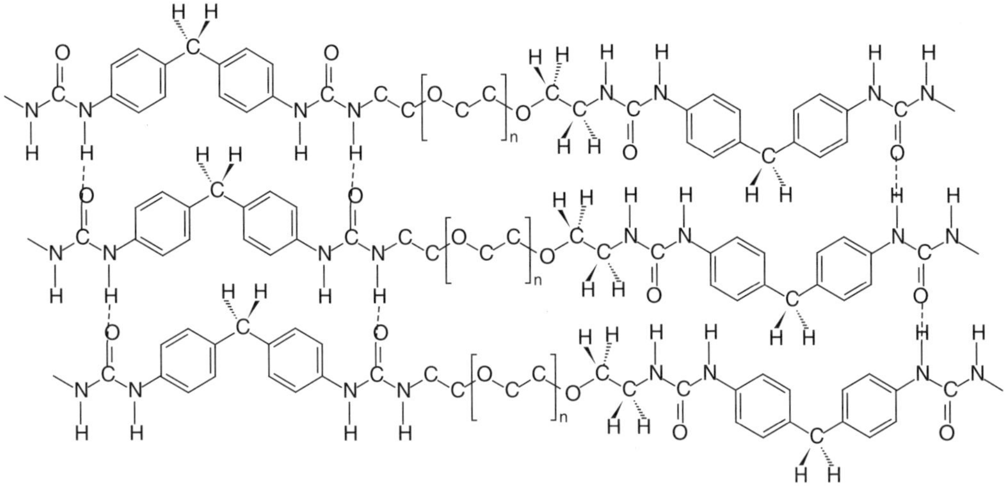

Figure 7.17 *Schematic of the structure of a polyurea.*

transition and then stays constant up to around 150 °C. The value of this modulus can be varied by altering the "hard" to "soft" block ratio by varying the amount of MDI and the molar mass of the poly(ethylene oxide). The combination of high modulus and flexibility is desirable in applications such as the sole of a shoe. This must support the load of the person, yet have the flexibility to deform as the person walks.

For a harder material, polyureas can be created by the reaction of an isocyanate with an amine, or can be created when water is present in the system (Figure 7.17).

7.6.2 Silicone elastomers

Silicone elastomers are encountered in a variety of applications, such as sealants used for baths and showers as well as other applications where gaps are to be filled with a water impervious flexible material, mouldings used in medical applications, fuser rolls in printers, etc. Silicone elastomers are usually based on polydimethylsiloxane (PDMS). Linear PDMS is normally a liquid at ambient temperatures. Even the very high molar mass materials show viscoelastic rather than solid behaviour. The polymer is a simple linear chain with terminal hydroxyl groups at each end. The T_g of the backbone is about −90 °C.

Elastomeric properties are obtained by lightly cross-linking the polymer chains. There are two types of rubber material: *room temperature vulcanised* (RTV) and *high temperature vulcanised* (HTV) polymers. The chemistry used to produce these elastomers is slightly different. For the RTV the cross-links are created by the reaction of the polymer with a reactive cross-linking agent, usually a hydrolysable tetrafunctional silane (Figure 7.18).

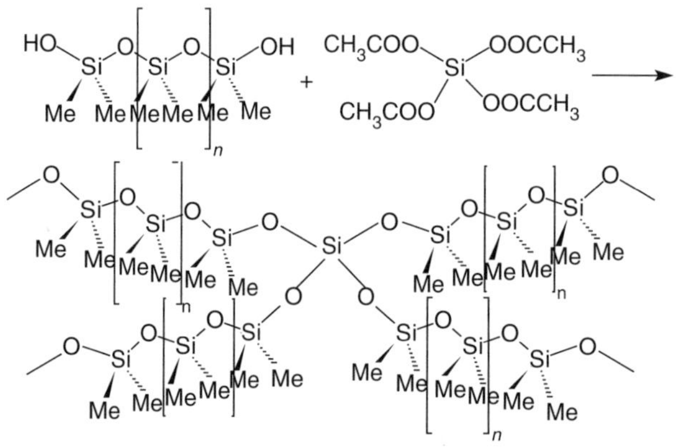

Figure 7.18 *Schematic of the RTV cross-linking process of PDMS to produce a siloxane rubber.*

The hydrolysis reaction creates silanol $[Si(OH)_4]$, which then rapidly condenses with terminal hydroxyl groups on the end of the PDMS to form a cross-linked structure. When the cross-linking agent is silicon tetraacetate $[Si(OCOCH_3)_4]$, the process will liberate acetic acid and these vapours can sometimes be detected as the RTV systems cure. An alternative cross-linking agent is tetraethoxysilane $[Si(C_2H_5O)_4; TEOS]$ and then the generation of the silanol liberates ethanol. This latter cross-linking agent is preferred as the smell is more acceptable than that of acetic acid.

The long silicon to oxygen bond and large distance between the neighbouring methyl groups makes the backbone very flexible with little steric hindrance to unit rotation. The polysiloxane chains can form a crystalline phase at about $-60\,^{\circ}C$. Nevertheless, the lightly cross-linked material still retains rubbery characteristics down to the T_g.

Lightly cross-linked PDMS elastomer can be further cross-linked and so converted into a hard solid. This is the basis of its use as a sealant.

A more stable matrix can be created using high temperature vulcanisation. The polymer used for this process contains a proportion of vinyl substituted silane units of the type shown in Figure 7.19.

The number of vinyl groups in a polymer chain can be varied and this will influence the physical properties of the material being created. The larger the number of vinyl groups, the greater the extent to which a carbon based cross-linked structure is created. Since the carbon based chain is resistant to hydrolysis, the HTV materials have better resistance to alkali attack than comparable RTV materials.

The HTV vinyl polymerisation is initiated using a platinum salt that is heated to a temperature above $100\,^{\circ}C$. The resulting matrix with both Si–O and C–C

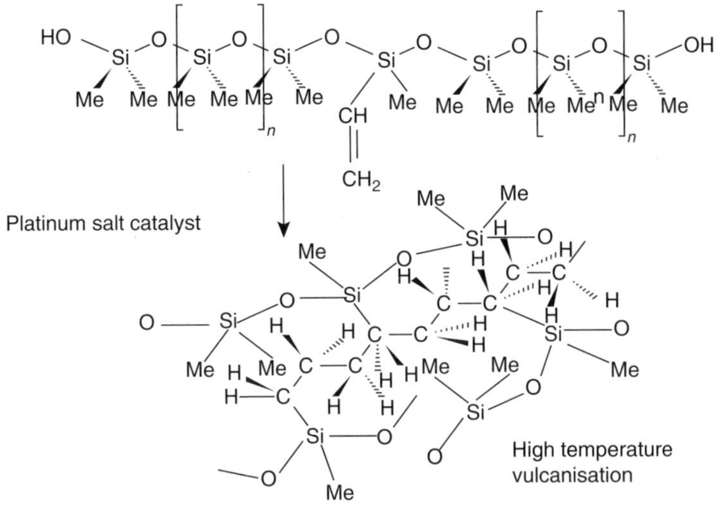

Figure 7.19 *Schematic of high temperature vulcanisation of vinyl substituted siloxanes.*

chains has very good thermal and chemical stability. Consequently HTV siloxane polymers are used in applications where operation to temperatures in excess of 150 °C may be desirable, for example as a rubber coating on fuser rollers in photocopiers.

Both RTV and HTV materials contain fumed silica (a non-crystalline, fine-grain form of silica with low density and high surface area) as an active reinforcing agent. The surface of the fumed silica contains Si–OH groups. These can undergo condensation reactions with silanol created by the hydrolysis process and so become incorporated into the cross-linked matrix. Without fillers silicone rubbers have very little tear strength and will readily snap if subjected to stress. However, adding 5–35% of fumed silica dramatically increases the tear strength, as well as the modulus, so making the material very useful.

7.7 Creep

The word "creep" means literally a slow movement. In everyday life this is usually from one point to another. However, in polymer technology the word is applied in two rather different situations. If a high modulus polymer is instantaneously extended to a given strain, the stress that is engendered slowly reduces. This relaxation of the stress is called *creep* and the elements of stress relaxation are discussed further in Chapter 10. If a stress is applied to a low modulus rubber, the material quickly extends to a length where the internal restoring force balances the external stress as described by the Mooney–Rivlin equation. However, if the stress is held constant, over a longer time the rubber may slowly extend further.

This slow second extension is also called creep. In both cases the strain associated with the creep may, or may not, be recoverable. However, whether or not the process is reversible, it must involve molecular movement of some sort.

The creep of a polymer not too far below its glass transition is ascribed to very slow chain movement, induced by the applied stress, of the kind displayed in the stretching of rubber, perhaps also with some change in the geometry of interchain entanglements as occurs in flow. This latter was introduced as reptation in Chapter 3 and will be described more fully in Chapter 8. If the strain is low, after removal of the external stress, the sample slowly returns to its unperturbed state as the molecules and the entanglements struggle to return to their equilibrium conformations.

The creep of stretched rubber is rather more complex. First, if the rubber chains are not all connected to each other through cross-links, the least tied chains may move further by escaping from, or rearranging, the entanglements that prevent short time translation. Such creep is irreversible. On the other hand, if all the chains are effectively linked to each other, albeit with only a few cross-links, there may still be escape from or rearrangement of the entanglements that hold the chains together like short-lived cross-links. On removal of the stress, the covalent cross-links pull the chains back to their original equilibrium arrangement and this creep is recoverable. This rearrangement of entanglements is treated theoretically using the "snake in a tube" model referred to earlier.

Finally, as rubber ages, further slower creep may take place as covalent bonds in the chains are broken by degradation reactions, a process that can be hastened when the rubber is under stress.

Further reading

Gedde U.W. *Polymer Physics*, Chapman and Hall, London, 1995.

8

The liquid/melt state

At high temperatures, non-cross-linked molecules start to move relative to each other, and creep or flow result. This is of enormous technological significance because this is the means by which plastic and rubber articles are processed into their final shape.

8.1 Viscosity

Polymer liquids and solutions are able to exhibit wide variations of viscosity. Low molar mass polymers can have viscosities which resemble those of mobile liquids like water, whereas higher molar mass materials of the same polymer type can have viscosities that resemble treacle or gum. This ability to achieve large variations in viscosity by changing the molar mass is a property of long chain polymeric materials. Indeed, this characteristic is of vital importance in all sorts of technological applications such as polymer processing, lubrication, coatings and adhesion. We all know high viscosity when we see it, but, in scientific terms, what exactly is it?

The first and the most easily understood definition was given by Newton as simply the resistance to flow. Newton showed that the shearing stress necessary to cause flow was proportional to the flow rate, with a constant of proportionality called the *viscosity coefficient*, given the symbol η:

$$\sigma = \eta \left(\frac{\partial \gamma}{\partial t} \right)$$

where $\partial \gamma$ is the change of the shearing strain in the time interval ∂t and σ is the measured stress. This is analogous to Hooke's law, except that stress and strain are shear quantities, strain rate $(\partial \gamma / \partial t)$ replaces strain, and the viscosity coefficient replaces the elastic modulus. We shall use Newton's viscosity coefficient extensively in what follows (Figure 8.1).

An alternative concept that is very relevant to our studies is that viscosity is a measure of the energy dissipated in flow. The more viscous a fluid is, the more work (force × distance) has to be done to make it flow a given distance, and so the more energy is dissipated. This definition will allow us to compare and contrast liquid viscosity with rubber energy damping.

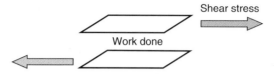

Figure 8.1 *Shearing a viscous liquid.*

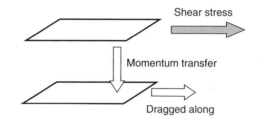

Figure 8.2 *Momentum transfer in shear.*

A third definition, which we shall not use, is that viscosity measures the momentum transferred from one plane in the liquid to another plane under the influence of an applied stress. This second plane is dragged along after the first, but falls behind as the planes separate (Figure 8.2).

Liquids which obey this simple relationship are described as being Newtonian. However, as will be shown later, many polymeric liquids are classed as non–Newtonian because they deviate from this simple relationship. The deviations are a direct consequence of the dynamic long chain nature of the molecules in the liquid. In simple liquids, such as toluene, kerosene or acetone, the movement of each molecule is essentially independent of the movement of another. However, in polymeric liquids, this usually is not correct.

When the viscosity of a polymer fluid is examined as a function of its molecular weight, or the viscosity of a polymer solution as a function of its concentration, we find two clear zones of viscosity behaviour (Figure 8.3).

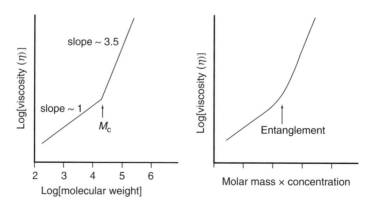

Figure 8.3 *Viscosity dependence on molar mass and concentration.*

In both cases, the viscosity starts to rise as the first power of molar mass or concentration, but at a critical point it becomes proportional to a higher (3.5 or 3.0) power. This critical point is the value at which the polymer chains start to entangle. Small chains cannot do so, and in dilute solution even long chains are separated. Short chains behave very much like simple molecules and their interactions with other molecules which surround them are relatively small. However, once the critical molar mass for entanglement has been exceeded, a mechanism for connectivity between the chains has been created and the higher power dependence is observed.

In the case of concentration, the plot of concentration multiplied by molar mass reflects the volume which the molecules occupy in solution, giving for each a *hydrodynamic volume*. In dilute solution, the long chains are sufficiently well separated that the hydrodynamic volumes of each do not overlap, and they can behave as isolated diffusing entities. As the concentration is increased, the hydrodynamic volumes will overlap and the chains will entangle, giving the dramatic increase in viscosity. For low molar mass polymers, the product of concentration multiplied by molar mass never reaches the critical value and the enhancement due to entanglement is not observed (Figure 8.4).

The very high viscosities of polymers arise because of the entanglements and the consequent great sensitivity to both molar mass and concentration.

So, "What are the molecules doing?" As was introduced in Chapter 3, the total movement is a combination of the internal rotation and translation occurring

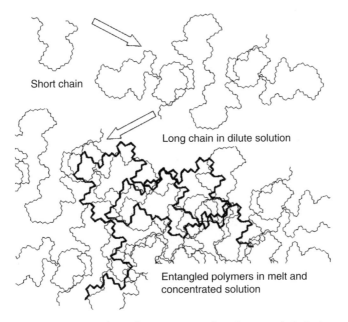

Short chain

Long chain in dilute solution

Entangled polymers in melt and concentrated solution

Figure 8.4 *Short, long, separated and entangled chains.*

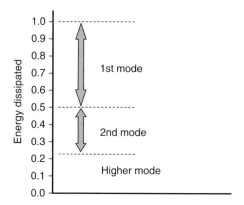

Figure 8.5 *Normal mode contributions to viscosity.*

as *reptation*. The transition from rubber to melt occurs when reptation has the energy and time to occur. But just how much time is needed for this composite motion? We saw that, because it has the highest energy barrier to overcome, it is the slowest of all the movements contributing to energy loss and viscosity.

When the individual chains are separated in dilute solution, the major movements absorbing energy are the normal modes of motion introduced in Chapter 3. At very low concentrations, the polymer molecules are able to act independently and the changes in viscosity that are observed with increasing shear rate are a consequence of the way that individual polymer chains respond to the external shearing stress. The first mode absorbs the most work and contributes most to the viscosity. In fact, the first mode contributes about half the viscosity due to these normal mode motions, while each higher mode contributes roughly half of the remainder (Figure 8.5).

When we looked at the time required for each normal mode of motion, we saw in Chapter 3 that the first mode required the most time, with the higher modes requiring progressively shorter times. In general terms, the time required for each mode is given, for the *i*th mode, by:

$$\tau_i = \frac{1.7(\eta - \eta_s)}{K_i ckT}$$

where η is the viscosity of the solution, η_s is the viscosity of the pure solvent, c is the solution concentration, k is Boltzmann's constant and K_i is a numerical constant for each mode, reflecting the lower contribution of the higher modes. The effect of molar mass is now incorporated in the solution viscosity.

For low molar mass polymers having no entanglements, the normal modes of motion of the polymer chain are the main contribution to the observed viscosity.

If the shear rate is higher than the time for the first normal mode, the chain does not have time to respond to the applied perturbation, and only the higher modes are able to be activated. In other words, at times which are shorter than τ_1 the first normal mode is frozen out and hence cannot contribute to the observed viscosity. Further increase in the rate of shear will progressively remove further modes until the viscosity falls to a value which corresponds to that of the solvent. This simple description, with minor modifications, describes the behaviour of most polymer molecules in dilute solution. Because in solution the backbone motions are effectively liberated, so that the chains are fully flexible, the description of the viscosity of dilute polymer solutions is essentially independent of the chemical nature of the molecules. The modes are purely defined by the end to end length of the polymer chains and hence by the molar mass of the polymer.

In summary, then, reptation is the prime cause of polymer liquid viscosity for polymers with molar mass above the critical entanglement value. The polymer chains are considered to be ideally flexible, the defining time being that required to escape from, then reform, entanglements.

8.2   Effect of shear rate

We turn now to the effect of time on liquid viscosity. To explore this, it is necessary to examine the effect of shear rate, essentially reciprocal time. In this context, of importance are the rates, or time constants, of the various energy absorbing processes that lead to the observed viscosity. Molecules need time to reptate and to undergo the normal modes of motion. If they are not given enough time then the movements cannot take place, no work is done, and so there is no contribution to the viscosity. Consequently, as the shear rate is increased, first the slowest motion, reptation, becomes impossible, then the first mode, and at higher shear rates the higher modes, until none of these motions contribute to the viscosity (Figure 8.6).

As the viscosity falls with increasing shear rate, does this mean that the polymer flows more easily? No; just the reverse, in fact. Remember that when a rubber

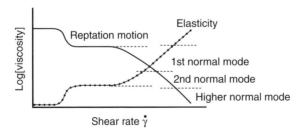

Figure 8.6   *Effect of shear rate on energy absorption by normal modes.*

was not given time for the internal rotation motion to take place, it moved back up the modulus curve and exhibited the properties of a glass. Similarly, when a fluid is not given the time necessary for the translation motions, it moves back up the modulus curve and exhibits the properties of a rubber. So, as the shear rate increases, the polymer becomes more and more rubber-like, showing increased elasticity and decreased energy absorption. At the highest shear rates, it behaves just like a rubber. This, too, is illustrated in Figure 8.6.

Since the sample has both viscosity and elasticity, it is said to be *viscoelastic*, and the phenomenon is called *viscoelasticity*. As the name implies, in this region the polymer liquid exhibits flow characteristics which are reflected in the term "viscosity" and rubber-like behaviour reflected in "elasticity".

A good example of this is provided by the children's toy called "bouncing putty". This is, in fact, liquid polydimethylsiloxane (PDMS), in which the polymer chain ends have hydroxyl groups, and the liquid is filled with solid boric oxide powder (Figure 8.7).

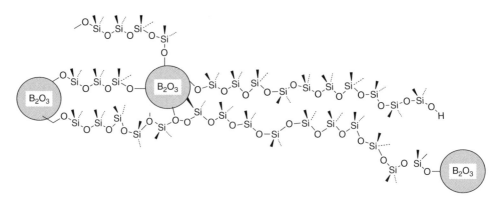

Figure 8.7 *"Bouncing putty"*.

The hydroxyl groups of the PDMS interact with the boric oxide to form a very weak unstable borate ester. The borate ester is sufficiently unstable for the bond to exist only on average about one second. Just as in reptation, the breaking of one bond allows another to be formed, and hence the average number of bonds remains approximately constant although they are continually breaking and remaking. While the bond is in existence, the chains cannot translate, and the material has the properties of a rubber. However, when the weak ester bond breaks, the chains can separate and the material has the properties of a liquid. So liquid behaviour takes place only in times longer than a second (Figure 8.8).

If the material is placed on a flat surface, it slowly flows to cover the surface. However, if it is moulded into a spherical shape and thrown on the floor, the impact time is only a fraction of a second, and the material bounces like a rubber ball.

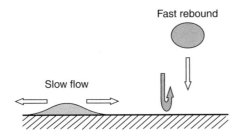

Figure 8.8 *Flow and bounce of "bouncing putty".*

8.3 Viscoelasticity in technology

We now examine two examples of the consequences of this viscoelasticity in technology. In both cases, the phenomena are concerned with the elasticity that arises at high shear rates, sometimes called *"elastic memory"*.

8.3.1 Die swell

During the extrusion of a molten polymer material, the process economics require the action to be carried out at the highest possible speed. However, if we go to too high extrusion rates, a deformed extrudate emerges, a phenomenon known as *"die swell"* (Figure 8.9).

When the polymer is forced through the die at high speed, there is not time for full flow to take place, and the elasticity makes the extrudate spring back towards its original thickness. This can be overcome by increasing the melt temperature so as to speed up the flow molecular motions. Of course, the temperature cannot be raised beyond the point at which the polymer starts to decompose.

8.3.2 Melt fracture

A direct consequence of the viscoelasticity of the melt is the phenomenon of melt fracture. If the liquid is subjected to high shear stresses then the polymer coils can become extended in the direction of the shear. The result of this distortion of the polymer coil will be to increase the number of entanglements which the chain experiences. In other words, the higher the rate at which the

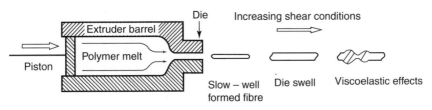

Figure 8.9 *Die swell at high extrusion rates.*

melt is sheared, the greater the number of entanglements. However, the viscosity depends directly on the number of entanglements and hence the "effective" viscosity depends on the magnitude of the shear stress. Once the liquid has become more entangled, it will take a finite time for it to recover its "equilibrium" number of entanglements, and this will provide a source of instability in the flow. The direct consequence of this is that fibres extruded under high shear rates can have very rough surfaces compared with those produced at slow extruder rates, which may be smooth. In the extreme, the entanglements can cause the liquid when it leaves the die to contract and the filaments to break!

8.3.3 Viscostatic lubricating oils

Modern lubricating oils, and especially those in internal combustion engines, have to operate over a wide range of temperatures. In North America cars may be required to operate on start up between temperatures as low as $-20\,°C$ in the winter and up to $40\,°C$ in the summer. It is desirable that the viscosity changes as little as possible over the working temperature range (hence the name viscostatic). This is often achieved by adding a polymer such as polyisobutylene, atactic polypropylene or a polyacrylate to the base oil. So viscostatic lubricating oil, being formulated from base oil and a polymer additive, is a polymer solution.

In order to allow the lubricant to work at low temperatures, the viscosity of the base oil has to be kept quite low. Also, the polymer molecules are coiled up, or even aggregated into a colloid, so they do not contribute to, or cause, an overly high viscosity. Then, as the temperature is increased, so the viscosity of the base oil will drop and its ability effectively to lubricate the bearings will decrease.

However, in shear at higher temperatures, the polymer molecules can be extended and give a viscosity increment that stops the viscosity falling too low. This stretching process is reflected in what is called the *elongational viscosity*. The application of the shear stretches the polymer coil and increases coil overlap and entanglement, which consequently increases the viscosity. So, contrary to what was presented in section 8.2, solutions close to the point where entanglement becomes possible can exhibit lower values of viscosity at low shear rates than at high shear rates.

Other mechanisms, such as using a polymer which separates into a colloid suspension at very low temperatures, but dissolves to exert an added viscosity at high temperatures, formation of hydrogen bonds, or creation of ionic interactions, are used to maintain polymer–polymer interactions and thereby sustain the effectiveness of the lubricant.

So, a properly formulated lubricant shows the desired properties of flow and energy absorption at the shear rates for which it has been designed. However, at higher shear rates, the polymer molecules do not have time to undergo the required movements, and the viscous fluid lubricating film behaves like a rubber membrane.

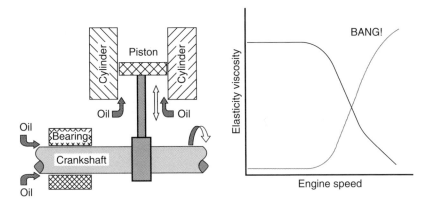

Figure 8.10 *Lubrication failure at high engine speeds.*

Consider, as an example, the representation of a bearing between a crankshaft and a piston in an internal combustion engine. The liquid lubricant is forced as a film between the moving surfaces. At reasonable engine speeds, the oil film flows between the bearing surfaces and does its job of lubrication, keeping the surfaces apart as it does so. However, at too high engine speeds, the lubricating film is subjected to very high shear rates, the liquid viscosity properties disappear and are replaced by a film that behaves like a rubber. The rubber membrane tears, and the surfaces come into contact, with disastrous effect (Figure 8.10).

8.3.4 Shear thickening and thinning effects

It is not uncommon to observe a situation where a slowly stirred polymer solution will have an apparent low viscosity. However, if the liquid is now subjected to a sustained shear field then the liquid will exhibit an increased viscosity. This phenomenon, called *shear thickening*, is caused by elongational viscosity creating interactions which do not exist in the equilibrium polymer solution.

In contrast, some polymers when allowed to rest may form larger clusters of molecular aggregates giving the solution the form of a gel. On shearing, they can then disaggregate and behave like isolated polymer molecules, so the solution becomes a mobile liquid. This type of phenomenon can be observed when the concentration of the polymer solution is close to the critical value for entanglement. There is a critical shear rate at which this disaggregation occurs and this is termed the *yield stress* in *shear thinning*.

A closely related phenomenon is *thixotropy*. This is the property of some polymer solutions to show a time dependent reduction in viscosity when sheared. The longer the fluid is subjected to the shear stress, the lower its viscosity becomes. So a thixotropic fluid is one which takes a finite time to attain its equilibrium viscosity in shear after being subjected to a step change in shear rate. In other words, for the shear thinning effect the independent variable is shear rate, whereas for the thixotropic effect it is shear time.

These two effects are particularly important in what are called *thixotropic* paints or coatings. In the can the paint is a semi-solid gel, but when stirred for a time and then applied by a brush or roller it spreads like a liquid, reverting to the semi-solid consistency when the applicator is removed.

So all these technological effects are controlled by the rate at which the polymer chains are able to execute movement and rearrangement in response to external stresses.

Further reading

Bird R. Byron, Curtiss C.F., Armstrong R.C. and Hassager O. *Dynamics of Polymeric Liquids*, 2nd edn., John Wiley & Sons, New York, 1987.

Furukawa J. *Physical Chemistry of Polymer Rheology*, Springer, Heidelberg, 2003.

9
Drawing and fracture

9.1 Introduction

Polymers are often subjected to different kinds of stress, either during a shaping process or during service life. They respond differently to the applied stress, depending on the molecular motion at that particular temperature. When a polymer is amorphous, at one end of the spectrum, where molecular motion occurs freely at room temperature, the polymer will behave as a rubber. It can be stretched to many time its original length without breaking. When the applied stress is released, it will go back to its original length. At the other end of the spectrum, where molecular motion is very restricted, the polymer is a glass and only deforms very slightly and then breaks.

If semi-crystalline polymers are stretched, a very different situation will be observed. When the polymer is elongated to a certain point, called the *yield point*, some parts of the material will be permanently elongated, while some other parts seem not to change at all. Since drawing can be carried out at room temperature, as opposed to high temperature, the process is called *cold drawing*. Drawing of polymeric materials is widely used to shape samples into long thin articles or to create fibres. Depending on the conditions, a plastic object may be drawn right down to a fibre, or it may be extended only up to a certain point when it breaks. This breakage is fracture in tension.

More generally, fracture is a vitally important characteristic of materials and is encountered in a variety of different circumstances. In order to determine the strength or resistance to breakage in use, a number of test methods are employed. In these, testing to destruction usually leads to failure by way of fracture of the specimen. The two methods most often used are *tensile testing* and *impact testing*. In the former, the material is subjected to a continuous tensile stress. Then the stress and the resulting strain are monitored until the sample breaks.

However, in the use of plastic materials fracture is often caused by a sudden impact, sometimes by a sharp object. The impact is then focussed at a point and the response of the material is reflected in the way in which the energy associated with the impact is dissipated. If the impact is able to initiate a crack then failure will be a consequence of the propagation of that crack through the sample. So impact test procedures have been developed to measure this.

In both tensile stress and impact, the ultimate fracture properties of a plastic are related to the ways in which the material is able to deform and absorb energy prior to the crack propagation that eventually leads to failure. Crystalline and amorphous polymers fail in slightly different ways and so will be considered separately. However, before examining the failure mechanisms, it is appropriate to say a word about these two test procedures.

9.2   Tensile testing

A widely used test method for measuring both drawing and fracture involves clamping two ends of a sample, applying an elongation stress, and then monitoring both the applied stress and the resulting strain. This usually observes the ability of the material to deform under constant rate of strain. However, often the change in cross-sectional area as the sample stretches is not measured, and so the test evaluates the nominal stress as was defined in Chapter 4. It is worthy of note that often the test equipment can measure the behaviour of the sample under application of a periodic stress, as will be discussed in Chapter 10. While the application of a steady strain rate appears to be measuring a static property, for polymeric materials the observation is influenced by the dynamic behaviour of the molecules concerned. So it is necessary to consider the various complex time dependent processes that are taking place. This can be done by making the measurements over a range of temperatures and strain rates.

9.3   Impact testing

If a piece of window glass is hit with a hammer, it will fracture into a number of fragments reflecting the strength and location of the impact. If we examine, using electron or optical microscopy, the surface resulting from the impact, we find that it is smooth. This indicates that the deformation has been focussed into a very narrow region and the energy associated with that deformation has propagated rapidly through the material. If, in contrast, we examine the failure surface of many plastics after impact, we find that the surface is rough, indicating that significant amounts of material have been torn out of the new faces of the sample as they have formed.

Impact testing equipment utilises either a dropping weight or a swinging pendulum. These are instrumented to monitor the load applied as a function of time and/or the specimen deflection prior to fracture. Modern sensing equipment uses digital photography, laser tracking, and strain gauges to do this. A complete picture of the failure process may require the use of different rates of impact, achieved by varying the height and weights of the impacting load. To make the tests realistic, it is also useful to select the shape of the impact probe to mimic the type of impact incident which the material will experience in use. The damage created by a needle shaped load will be very different from that caused by a ball or cylindrical shaped weight. As an example in use, smooth rocks travelling down a plastic pipe will create a very different impact from a blow by a sharp needle shaped object. Similarly, a spanner falling on an aircraft wing will have a different effect from a bird impacting the same structure at high speed. In attempting to model such events, it is often difficult to quantify the effects of the impact of falling or propelled objects, especially for composite structures, where damage may occur within the mass of the material and not necessarily at the surface.

In reality, most impacts are biaxial rather than unidirectional and are further complicated by the fracture having varying degrees of brittle or ductile characteristics. Brittle materials, like un-toughened window glass, require little energy to start to crack and little more to propagate the crack to a shattering climax. Other materials possess varying degrees of ductility. Highly ductile materials fail in drop weight testing by being punctured and require a high energy load to initiate and propagate the breakage.

Many materials are capable of either ductile or brittle failure, depending on the rate and temperature conditions used, and can then exhibit a brittle to ductile transition with increasing temperature or decreasing strain rate. The area under the stress–strain curve (equivalent to force × distance) gives the energy required to achieve failure. In an impact test, this fracture energy is called the *impact energy* or *impact strength*. Another parameter which is measured is the *impact velocity* or *strain rate*. In some materials, dropping a 5 kg weight from 1 m produces a very different result compared with a 1 kg weight from 5 m, since the material exhibits strain rate sensitivity. Every material will behave differently, depending on the geometry of the impact surface, on how the specimen is clamped, and on the geometry of the clamping.

As opposed to window glass, in a plastic the energy associated with the deformation can often be dissipated in ways that do not lead to breaking of chemical bonds or cause similar interatomic disruption. As a consequence, there can be a greater resistance to failure and this gives certain plastics high impact strengths.

9.4 Drawing of amorphous polymers

In what follows, *nominal stress* is defined as the load on the sample, registered in tension, divided by the initial cross-sectional area of the sample. *True stress* is the load divided by the actual cross-section of the sample as it narrows under extension. When a sample is subjected to a tensile force or load at a constant rate of strain, the stress measured as a function of strain can show certain unusual features (Figure 9.1).

There are four regions of different stress–strain behavior. In region 1, the polymer experiences elastic deformation and recoverable strain. In this region the molecules are deformed in shape as much as the available molecular motions

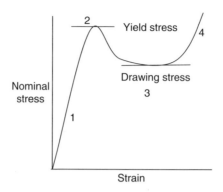

Figure 9.1 *Nominal stress as a function of strain for a typical amorphous polymer.*

will permit. Then, again if molecular motions permit, in region 2 the nominal stress goes through a maximum. The sample shows the clear development of a narrow neck. If, instead of nominal stress, the true stress is calculated, in region 2 the sample shows a change in modulus (slope of the stress–strain curve), but sometimes not a maximum (Figure 9.2).

Nonetheless, in either representation there is a decrease in modulus at this point. This is called the *yield point* or *yield stress* of the polymer. At some point of weakness a *neck* forms and further change starts to occur at the neck. Several theories have been put forward to explain this yield point. In amorphous polymers it is assumed that in the initial elastic recovery region, 1, the molecular conformation change is small and does not cross any large energy barrier. However, when the stress approaches the yield point, the conformation is sufficiently distorted to cross a barrier into a state from which it cannot spontaneously return. In this state, further deformation is easier than in the original relaxed state and the chains can migrate to their new position in the neck. It has been suggested that this molecular translation process is aided by the temperature rise

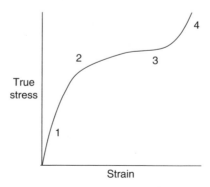

Figure 9.2 *True stress as a function of strain for an amorphous polymer.*

caused by adiabatic absorption of the stress–strain energy. Certainly this seems to be a possibility when a sample is cold drawn at high shear rates. Whatever is actually happening, the real modulus is less than that of the unperturbed polymer.

In region 3 of Figures 9.1 and 9.2, the nominal stress is a broad minimum as the sample elongates and the neck narrows until it breaks. Here there is some evidence of an adiabatic temperature effect because the drawing stress and modulus are observed to fall if an already high shear rate is increased even further.

As this approach to fracture is happening, in region 4 of the figures the stress rises again. This property is known as *strain hardening*. In this region, the chains in the remainder of the neck are becoming excessively deformed with a considerable gain of conformational energy and loss of conformational entropy, and so oppose further movement.

9.5 Drawing of crystalline polymers

Drawing is one of the most important industrial processes for producing polymeric fibres. The large permanent deformation evidenced in drawing cannot be obtained without some kind of molecular motion. Indeed, various molecular processes occur in cold drawing of semi-crystalline polymers. These include breaking of lamellae into smaller connected blocks, rearrangement and aligning of these structural blocks in the drawing direction, chain straightening by unfolding of lamellae, sliding of molecules passing one another, and, of course, breaking of molecular chains.

Again, in crystalline and semi-crystalline polymers there are four regions of different stress–strain behavior. As before, in region 1 the polymer experiences elastic deformation and recoverable strain. In this region, the spatial arrangement of the crystal lamellae within spherulites becomes deformed as much as the reversible molecular conformation change of the inter-crystalline tie molecules will permit.

In region 2, the nominal stress goes through a maximum which may involve two yield points. The first yield point is ascribed to a recoverable orientation of the lamellae within the spherulites. In this region, the lamellar crystallites are pulled into an orientation parallel to the direction of tensile stress. In fact, the final average orientation is about 45° to the draw direction. Molecules will change from a random orientation in the isotropic starting material to a preferred orientation. The sample is therefore anisotropic and stronger in the drawing direction (Figure 9.3).

The second yield point is caused by irreversible fracturing of the lamellae and spherulites into a fibrillar structure (Figure 9.4). This second yield point is often not clearly visible in the stress–strain data, but is evidenced in a change in the X-ray scattering pattern. The change in the crystal structure as the lamellae

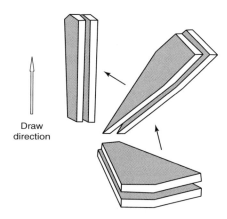

Figure 9.3   *Orientation of lamellae on drawing.*

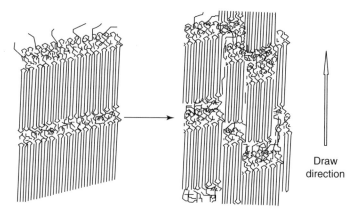

Figure 9.4   *Fibril formation and martensitic transformation.*

become fibrillar is often referred to as a *martensitic transition*. Significantly, as the sample passes through the second yield point, it starts to show the clear development of a narrow neck at one point in the sample.

As the strain continues, the neck in the sample grows in length but does not change in thickness (region 3). This is called *stable necking* and the load applied is called the *drawing stress*. This drawing continues until the neck reaches the clamps holding the sample (Figure 9.5).

If the original sample has the normal test geometry, the neck takes the form of a tape, as illustrated. However, if the sample is an extrudate like a thick fibre, the drawn neck is a very thin, but very strong, high modulus fibre.

This fibrillation requires considerable conformational and translational movement of the tie molecules and involves restructuring of the entanglement points that hold these chains together. Just as in the creep of rubbers and flow of melts, once the process of entanglement rearrangement has started, it becomes easier, resulting

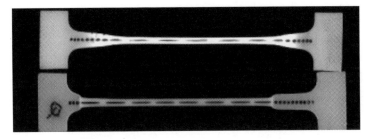

Figure 9.5 *Neck formation at stage 3, in the stress strain curve figure 9.2.*

in a decrease in nominal modulus in solids and lower molar mass dependence in viscous flow. This process by which chain movement allows rotation and then thinning of the lamellae is called *chain slip* in drawing technology, and corresponds to the rearrangement of the containing tube in the theoretical models of reptation.

In crystalline and semi-crystalline polymers, too, it has been suggested that this molecular rearrangement involving both conformation change and entanglement migration is aided by the temperature rise caused by adiabatic absorption of the stress–strain energy.

A polymer cannot be drawn into a fibre at strains below that of the second yield point. If we consider the ratio of the length of a small element of stretched polymer to the original length, we get what is called the *draw ratio*. Then the ratio in the neck as it starts to form, when the sample passes the second yield point, is called the *natural draw ratio*. This represents a minimum draw ratio below which a fibre cannot be drawn from the sample. It is not a constant, but depends on the characteristics of the polymer such as molar mass, crystallinity and pre-orientation.

9.6 High modulus fibres

Long polymer molecules are anisotropic in nature, i.e. their properties are direction dependent. If a single polymer chain is considered, properties along the chain direction are determined by covalent bonding, while those in the perpendicular direction are determined by van der Waals forces. However, on a scale that is much larger than the size of the molecules, this anisotropy is not observed. The anisotropic nature of the chain is averaged out because of the random arrangement of the chains in an amorphous polymer or of crystallites with a chain folded structure in semi-crystalline polymers. If all chains are fully extended and aligned so that their chain axes are parallel to one another then the anisotropy can be expected at a macroscopic level. This was realised very early in the history of polymer science. So, very high theoretical moduli and strengths, many times greater than those observed in normal polymeric materials, were suggested.

The theoretical modulus and strength of a polymer crystal are determined by the shape of the molecules in the unit cell and the packing of chains in the cell. They are also determined by secondary interchain interactions in the direction perpendicular to the chain direction. As such, polymers with a helical structure tend to have rather low theoretical moduli, since the force constants for bond twisting are much lower than for bond opening and bond stretching. An example of this can be seen by comparing polyethylene and polypropylene. The former has a planar zigzag structure, while the latter has a helical structure. As a result, the theoretical modulus for polyethylene is calculated to be as high as 180 GPa, while that of polypropylene is around 40 GPa.

Since polymer chains contain a large number of single bonds, which can be rotated around the bond axis, they are said to be flexible and tend to fold back and forth on themselves when they crystallise. To obtain an idealised oriented structure, these chains have to be unfolded and aligned in a particular direction. This can be achieved by drawing under appropriate conditions. We saw in section 9.5 that semi-crystalline polymers can be cold drawn. The mechanical strengths of these drawn polymers are significantly greater than those of the original isotropic structures. The change is due mainly to the molecular alignment and chain extended structure in the drawn polymers, which are said to be *oriented*.

Polyethylene is the most studied polymer for the production of high modulus and high strength materials. It has been shown that the modulus achieved depends on the draw ratio, i.e. the higher the draw ratio, the stiffer the polymer. For high density polyethylene that can be normally melt processed, the modulus and strength achieved are around 40 GPa and 1.0 GPa, respectively. These are limited by two factors: high entanglement density and low molar mass. Entanglement prevents drawing to a very high draw ratio, while low molar mass limits the strength. However, with present spinning technology of ultra high molecular weight polyethylene (UHMWPE), fibres with moduli in excess of 100 GPa and strengths more than 2.5 GPa can easily be produced on a large scale. These fibres are many times stronger than steel on a weight-for-weight basis. The commercial products are known as Spectra® and Dyneema® fibres.

Another method for the production of high stiffness and high strength polymeric fibres is to prevent chain folding in the first place. One example of a structure which is introduced into the polymer main chain to prevent folding is a bulky aromatic group such as a benzene ring or biphenyl. This creates a rigid rod molecular structure which can form a liquid crystalline state. This was the method that in the 1970s first led to the commercial production of a high performance fibre, an aramid commonly known as Kevlar.

9.7   Failure in amorphous plastics: brittle and ductile fracture

The neck cross-sectional area remains unaltered during drawing until the neck has grown in length to reach the clamps holding the sample. Then a complex failure mechanism can occur by either of two processes: brittle fracture and ductile fracture.

If we cool either an amorphous plastic such as polystyrene or polycarbonate, or an elastomer such as natural rubber, to very low temperatures, the failure surfaces created on sudden impact are essentially the same as those observed from window glass. These, as seen by optical or electron microscopy, are very smooth, indicating that the failure progressed very rapidly through the material. So at low temperatures the polymer material is a brittle glass showing a high modulus, as indicated by the steep initial slope of the stress against strain curve A in Figure 9.6.

The impact has propagated in a very narrow zone and chemical bonds have been broken. If we examine a diagram of applied stress against resulting strain, the area under the curve, which measures the energy dissipated or work done in the process (work is force times distance), is relatively small. Furthermore, this figure shows that failure occurs after only a small extension of the material.

Increasing the temperature of the material causes a small drop in the modulus. This is the β-transition described in Chapter 5. It results in an increase in the yield prior to failure. Then the area under the stress–strain curve increases as shown in Figure 9.6, curve B. This change in the nature of the fracture is associated with the onset of local motions of the chain, even though they do not involve total internal rotation of the chain backbone.

At first sight, it is puzzling why the onset of such restricted motions should influence failure properties that are normally considered to involve the large scale movement of polymer chains. The explanation lies in the way in which

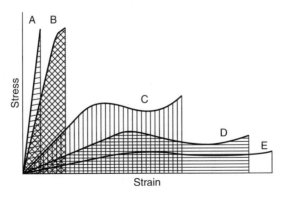

Figure 9.6   *Typical stress–strain plots for different types of failure: brittle (A); brittle–ductile (B); ductile–brittle (C); ductile (D); and rubber (E).*

the deformation propagates in the material. For bonds to be broken, the energy must be constrained in a small zone. If the energy can be effectively spread over a larger zone then it will be defocussed and more will be required to achieve failure. If the molecules are able to absorb energy in the β-processes then the energy can be dissipated away from the immediate impact zone. At these temperatures, the stress–strain behaviour shows a rate dependence, which illustrates the spread of the energy via the β-relaxation process. At very low or very high loading rates, the failure occurs at low strain. The highest strain to failure is observed when the loading rate corresponds to the β-peak in the loss against strain rate curve at the temperature of the test. So the stress–strain–fracture characteristics change to an increasing character of ductile fracture.

As the temperature is increased from that of the β-transition towards that of the main glass to rubber transition, the failure shows more and more ductile character (curve C of Figure 9.6). Since polymers are thermal insulators, the energy released as heat will be retained close to the crack. Then local heating processes can lead to expansion of the polymer lattice and so enable the cooperative molecular motions of the main glass to rubber transition.

The activation of the cooperative T_g modes is coupled with material being pulled out of the solid under the influence of the high stresses which are active close to failure. This, of course, gives the fracture surfaces their rough appearance. This will be discussed in more detail in the section on crazes (section 9.8). The initial slope of the stress–strain curve decreases, reflecting the lower modulus. At the same time, the matrix will undergo a greater degree of elastic distortion, as seen in Figure 9.6, curve C. A characteristic of this yield behaviour is the reduction in observed stress as the chains move before the ultimate failure occurs.

As the T_g is approached, so the initial slope decreases even further, again reflecting the lower modulus. The incipient molecular motion gives the matrix the ability to absorb energy. As a result, the energy that has to be put in to cause failure will also be increased. The larger area under the stress–strain curve in Figure 9.6, curve D indicates this. Further increasing the temperature allows the matrix to become more elastic and a high degree of strain is achievable before failure is achieved. The degree of elastic deformation before failure will depend on the molar mass of the polymer. If the molar mass is just above M_c, the critical entanglement molecular weight discussed in Chapter 7, the extent to which the matrix can be stretched before failure will be limited. For molar masses well above M_c, the entanglements will allow a significant degree of elastic distortion before failure occurs.

The curves of the storage and loss moduli for an idealised linear amorphous polymer are shown once more in Figure 9.7, this time with the five failure regions indicated.

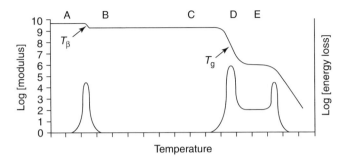

Figure 9.7 *Temperature dependence of modulus and loss for a linear amorphous*

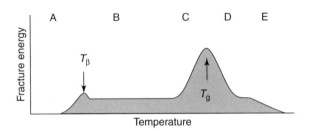

Figure 9.8 *Plot of variation of fracture energy with temperature.*

In the rubbery plateau region (curve E in Figure 9.6 and area E in Figure 9.7), the elastic distortion will be fully determined by the degree of entanglement.

Overall then, the fracture toughness, or fracture energy, which is the integral under the stress–strain curve, changes with temperature and shows a maximum at the T_g as indicated in Figure 9.8.

The strength of a plastic can be related, first, to the number average molecular weight of the polymer using a simple formula presented by Flory more than 60 years ago:

$$\sigma_w = A - (B/M_n)$$

where σ_w is the weighted average of the tensile strengths of the individual molecular weights. At first sight this equation looks a little strange since it equates the weight average of a property to the number average of a second property. Nevertheless, the equation has proved very useful in predicting the tensile strength of polymer blends. In this context, low molecular weight polymer is often blended with high molecular weight material in order to improve the ease of processing.

The ultimate (breaking) stresses can also be inserted into the empirical Williams–Landel–Ferry (WLF) equation introduced in Chapter 4, here using the shift factors, constants and with the glass transition temperatures, T_g, as the reference temperatures T_0:

$$\log\left[\sigma_{\text{ultimate}}(T)\big/\sigma_{\text{ultimate}}(T_g)\right] = \left[a(T-T_g)\right]\big/\left[b+(T-T_g)\right]$$

Since the WLF equation can also represent the influence of free volume, we see that it is a wide reflection of the effect of the chain dynamics not only on the modulus, but also on the ultimate strength, of the material.

9.8 Cracking and crazing

Whether it is experiencing impact or tension, the polymer is attempting to respond to an applied load. In impact, the load is applied at a very high rate and often concentrated at a point (Figure 9.9).

In the region of the impact, the plastic will experience a shock wave. In brittle fracture at low temperatures, the high frequency displacement will cause the energy to be constrained in a narrow zone where the impact occurs. Since the polymer is rigid, no molecular motions to accommodate the stress are possible. Then all the energy breaks bonds or overcomes polymer–polymer interactions and so separates the polymer chains, causing the material to break and fail.

However, when a plastic is subjected to shock above its β-transition, the propagation of the shock wave will be modified and ductile fracture is observed (Figure 9.10). In this diagram the grey shaded areas indicate where the energy is dissipated from the initial crack and from its propagation though the material as failure occurs.

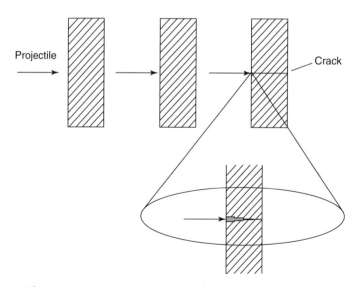

Figure 9.9 *Impact and subsequent crack propagation.*

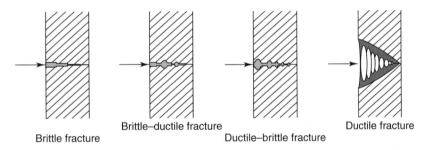

Brittle fracture

Brittle–ductile fracture

Ductile–brittle fracture

Ductile fracture

Figure 9.10 *Schematic of crack propagation for various states of the polymer.*

There are several important phenomena that are observed in ductile fracture. Initially, not far above the β-transition temperature, the polymer molecules are able to absorb energy and limited coupling of the side chain motion to the backbone, exemplified by the α and β relaxation processes discussed previously, will occur. The spreading of the energy is depicted by the increase in the shaded area. Initially the failure will be brittle–ductile, corresponding to a situation where the crack propagates rapidly. However, after a short period the energy is spread at right angles to the line of crack propagation, and dissipation occurs. What happens is that the loss associated with the β-process effectively heats the region of the matrix close to the crack. As the temperature is increased, movement of the polymer chains increases, and this reduces the energy available for crack growth. The crack is in effect arrested. Thus measurements of the crack propagation can be used to measure the fracture energy.

At higher temperatures the crack is unable to propagate without being arrested and the applied energy is converted into heat. Once the β-process has been activated, the fracture surface ceases to be glassy smooth, indicating the arrest of the propagating crack with material being pulled out of the separating surfaces as a consequence of the energy dissipation.

As the temperature is increased towards the glass transition, the energy dissipation becomes larger and the polymer is able to be deformed to a much larger extent before it fails. The macroscopic appearance of the crack surface becomes white, a phenomenon that is known as crazing. Microscopic examination shows that the crack surfaces have become very rough and covered by the remnants of strings of polymer pulled out to bridge the crack as it propagates. These "strings" are called *fibrils*. They are not completely separated threads, but are joined by smaller inter-fibrillar tie fibrils. It is the scattering of light by these fibrils that gives the white appearance of the craze. Under the influence of the breaking stress, these fibrils form narrow micro-necks and eventually fail.

Ductile failure is observed when it is possible for the polymer to exhibit rubber-like characteristics above the glass transition. So, the crazing by formation of the fibrous structure is controlled by both the value of the glass transition temperature and the viscoelastic characteristics of the rubber state.

As was shown in Chapter 5, the relevant motions of the polymer chains are controlled by the timescale of the stress application, by the available energy, and by the free volume in the system. We reiterate: in order to move, molecules need time, energy and space. So the response of the polymers, and their ability to absorb energy and move, determines whether the fracture is ductile or brittle, glassy smooth or accompanied by crazing and material "pull out".

Further reading

Brostow W. and Corneliussen D. *Failure in Plastics*, Hanser Publishers, Munich, 1986.

Ciferri A. and Ward I.M. *Ultra High Modulus Polymers*, Applied Science Publishers, Barking, 1979.

Ward I.M. and Sweeney J. *The Mechanical Properties of Polymers*, John Wiley & Sons, Chichester, 2008.

10

Dynamic mechanical relaxation

10.1   Periodic stress and strain

Although we have mentioned vibration damping, and have talked about frequency as the reciprocal of time, we have not yet made a proper examination of stress and strain that alternate in a periodic, sine wave, fashion. More specifically, we want to know in more detail how quantities like modulus and loss are related to each other, and how they vary with the frequency of the alternating stress.

10.2   Real and imaginary strain

We start by considering the stress as a function of time. This is represented as a sine wave in Figure 10.1. Then the strain is also a sine wave with the same frequency. However, for strain to be evidenced, the molecules must move, and this takes time. So the strain wave is delayed in time behind the stress wave, and thus there is a phase lag between the two waves as shown in Figure 10.1.

The two waves can be represented as two vectors, rotating in a space defined by the phase of the wave. This is called an *Argand*, or phase space, diagram. If the stress is represented by a vertical vector, the strain vector lags behind the stress by an angle determined by the time necessary for the molecules to move. This strain can then be resolved into two components, one *in phase* with the stress, and the other 90° *out of phase*. Now, wave functions are complex, in the mathematical sense, having real and imaginary components. The in-phase strain is the real component, and the 90° out-of-phase strain is the imaginary component. The complex strain, divided by the complex stress, leads to real and imaginary compliances, such that:

$$J^* = J' - iJ''$$

where J^* is the complex compliance, J' is the resolved real (in-phase with the stress) compliance, J'' is the resolved imaginary (out-of-phase with the stress) compliance and i is $\sqrt{-1}$. Since the real strain is in phase with the stress, it represents the movement of molecules that "keep up with" the stress, and so require no force, or work, to move them. Thus the in-phase compliance represents the elastic storage of energy and is called the *storage compliance*. On the other hand, the

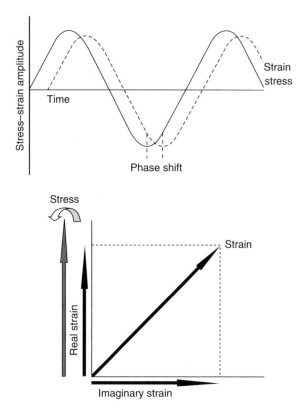

Figure 10.1   *Stress and strain as vectors in phase space.*

imaginary strain represents molecules that do not "keep up with" the stress. The molecules are coming when they should be going and they have work done on them by the applied force, which leads to an absorption of energy. As a result the out-of-phase compliance measures work done to force the molecules to move, measures the energy absorbed, and so is called the *loss compliance*.

Inversion of the complex functions leads to:

$$E^* = E' + iE''$$

where E^* is the complex modulus, E' is the real resolved modulus (*storage modulus*) and E'' is the imaginary resolved modulus (*loss modulus*). Note that inverting the equation has changed the sign. This first finding illustrates that the storage and loss moduli are not unconnected quantities, but are the real and imaginary resolved components of a single vector. We can use this concept of real and imaginary vectors to restate the dependence of loss on temperature and time as we move through a transition (Figure 10.2).

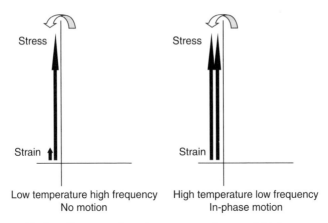

Figure 10.2 *Stress and strain vectors at low and high temperatures.*

Below the transition, where the molecules have neither enough energy nor enough time, there is little strain, so both the real and imaginary strain vectors are close to zero (high real modulus, zero loss). At the other extreme, when the molecules have plenty of energy and plenty of time, the strain can keep up with the stress, so that the real strain component is large (small real modulus) and the imaginary component is close to zero (zero loss). Then, again, the imaginary, loss component goes through a maximum at the transition temperature and time. In this case, of course, time is the time given in one cycle of the periodic stress, and so is the reciprocal of the periodic stress frequency. The complex modulus measured with periodic stress/strain is called the *dynamic mechanical modulus*.

10.3 Periodic shear: dynamic viscosity

Similar principles to the above apply when a liquid is subjected to an alternating shear stress. However, the measure of viscosity under shear is the rate of strain and not the strain itself, i.e. whereas Hooke's law said $\sigma = E\gamma$, Newton's law says that $\sigma = \eta(\partial\gamma/\partial t)$. Now, the phase of a differential of a periodic function *leads* the phase of the function itself (Figure 10.3).

In this case the real strain rate vector, giving the real viscosity, is 90° ahead in phase of the stress, while the imaginary strain rate vector is the one in phase with the stress. So, the complex viscosity is a combination of the two components, and with the imaginary component being in the negative phase quadrant ahead of the stress. Again, dividing the two complex functions gives:

$$\eta^* = \eta' - i\eta''$$

The complex viscosity is called the *dynamic viscosity*, and the loss of energy viscosity is the *real viscosity*, while the in-phase or elastic component is the *imaginary*

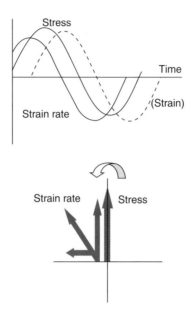

Figure 10.3 *Periodic shear stress and strain rate vectors.*

viscosity. Notice that the change from strain to strain rate results in the viscosity loss being real and the elasticity being imaginary, exactly the opposite of what we found when examining the modulus.

In everything we have done so far, we have drawn many graphs of modulus/loss against temperature/time/frequency, but nowhere did we have a specific equation for these curves. We shall now derive formulae for compliance and loss against time and frequency. The equations for temperature are more complex, but can be obtained quite simply from those for frequency by applying a frequency–temperature relationship such as the Arrhenius equation.

10.4 Definition of relaxation

Relaxation is a perfectly general phenomenon encountered in many different branches of physics, and even everyday life.

Consider three quite different phenomena, namely the cooling of a cup of hot coffee, a first-order chemical reaction, and the stretching of a piece of rubber under the influence of a constant stress. In each case, the selected object has been subjected to some change of condition: the hot coffee has been taken out of the water boiler and put in a cup, the chemical reagent has been taken out of a bottle in a refrigerator and put in a reaction vessel where it may have been dissolved in a solvent and heated, the unstressed rubber is clamped in holding jaws and subjected to a stretching force. Then the system under observation reacts to the change by moving towards a new state: the coffee cools, the reactant goes to products, the rubber stretches as the molecules deform.

Figure 10.4 *Three examples of relaxation.*

However, each reaction requires time. Then the observed property of the system finishes up with a new equilibrium value appropriate to the new circumstances. It is the time dependent approach to this new equilibrium value that is called *relaxation* (Figure 10.4).

So our definition is:

Relaxation is the time dependent return to equilibrium of a system that has experienced a change in the constraints acting on it.

The constraints may be familiar thermodynamic variables like temperature, pressure, volume, or they may be applied forces like stress or an electric voltage.

In every case, the return to the new equilibrium value proceeds at a rate that is proportional to the distance that the system is from equilibrium. So relaxation is described by a first-order linear differential equation, the solution to which is a simple exponential function.

For coffee:

$$-\left(\partial \Delta T/\partial t\right) = k_{N}\Delta T, \quad \Delta T = \left(\Delta T\right)_{0}\exp\left(-k_{N}t\right)$$

where k_N is the Newtonian cooling constant for the system. In other words, the temperature difference between the cup of coffee and its surroundings will approach zero in an exponential fashion.

For a first-order chemical reaction:

$$-\left(\partial c/\partial t\right) = k_c c, \quad c = c_0 \exp\left(k_c t\right)$$

where k_c is the first-order rate constant for the reaction.

For stretched rubber returning to zero strain:

$$\left(\partial \gamma/\partial t\right) = k\gamma, \quad \gamma = \gamma_0 \exp\left(-k_s t\right)$$

where k_s is a constant descriptive of the rate of movement.

All of these first-order linear differential equations and their exponential solutions can be generalised in an equation for the exponential decay of an observed property difference, P, from a value P_0 under the original set of constraints to zero after an instantaneous step-function change (up or down) in the constraints. We replace the rate constants, k, by $1/\tau$, where τ is a constant with the units of time and called the *relaxation time* of the system (Figure 10.5):

$$P = P_0 \exp(-t/\tau)$$

By setting $t = \tau$ we find that the relaxation time, τ, is the time required for P to fall to $1/e$ of P_0, where e is the Napierian logarithm base, 2.718.

This equation can be generalised as $P/P_0 = \varphi(t)$, where $\varphi(t)$ is the exponential time function, $\exp(t/\tau)$ is called the Debye equation. It should be noted that for all the many possible relaxation phenomena, the time function has only one adjustable constant, τ.

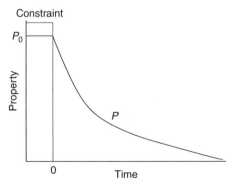

Figure 10.5 *Exponential return to new equilibrium.*

10.5   Relaxation as a function of frequency

If the observed property varies as a function of time in a way described by only a single constant, τ, and since the phase lag is a result of the time necessary for the system to adjust, it follows that the dependence on frequency should also be described by some equation, again with only one adjustable constant.

To find this equation, we rename the observed property difference, P, as the periodic response amplitude of the system, R. The property P can be stress, strain, pressure, temperature, dielectric permittivity, etc. The variation of the constraint and the response, in a completely general form, is illustrated in Figure 10.6. The figure is the same as for the stress and strain (Figure 10.1) and can represent the variation with time of any intrinsic property of the system P.

The following mathematical treatment is therefore completely general for any relaxation process. The mathematical technique for going from a function of a real variable, like time, to the corresponding variable, here frequency, is via the Fourier transform. However, the Fourier transform is "two sided", with both positive and

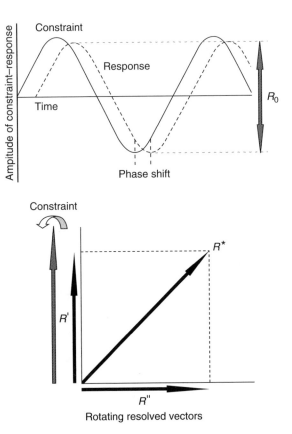

Figure 10.6   *Relaxation with periodic constraint.*

negative values of the real variable. But there is no such thing as negative time, so we use what is called a "one-sided Fourier transform", or the Laplace transform. Then the corresponding equation for the complex frequency dependent response is:

$$R^*/R_0 = L\left(-\partial\left(\varphi(t)\right)/\partial t\right)$$

where L represents the Laplace transform. Then, to find the equation for the complex response, all we have to do is differentiate the time function and look up the Laplace transform in a table of such formulae.

In fact, the result for a response as above is the amazingly simple equation:

$$R^*/R_0 = 1/(1 \pm i\omega\tau)$$

where ω is the frequency and τ is the self-same relaxation time. We can rationalise this equation by multiplying numerator and denominator by the complex conjugate of the denominator. For example, consider $1/(1+i\omega\tau)$ (as in compliance).

$$R^*/R_0 = (1-i\omega\tau)/(1+i\omega\tau)(1-i\omega\tau)$$

$$= 1/\left(1+\omega^2\tau^2\right) - i\omega\tau/\left(1+\omega^2\tau^2\right)$$

$$= R'/R_0 - iR''/R_0$$

So the real response, relative to R_0 and as a function of frequency, is given by the formula $1/\left(1+\omega^2\tau^2\right)$, which is our familiar sigmoid curve running from 1 when $\omega\tau$ is very much less than 1, through a point of inflection at 0.5 when $\omega\tau = 1$, to 0 when $\omega\tau$ is very much greater than 1. The imaginary response, R''/R_0, runs from 0 when $\omega\tau$ is much less than 1, through a maximum value of 0.5 when $\omega\tau = 1$, to 0 again when $\omega\tau$ is much greater than 1. This is the complete bell-shaped curve for our loss function.

10.6   Frequency dependent compliance

We are now in a position to specify completely the behaviour of both the energy storage (real) and the energy loss (imaginary) properties as a function of frequency. Combining the complex stress constraint and complex strain response, the overall response property is the compliance, J^*. Using the response equation above, and setting the glass compliance to zero gives

$$J^*(\omega)/J_0 = J'/J_0 - iJ''/J_0$$

$$= 1/\left(1+\omega^2\tau^2\right) - i\omega\tau/\left(1+\omega^2\tau^2\right)$$

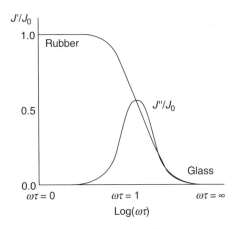

Figure 10.7 *Real and imaginary compliance as a function of frequency.*

Here τ is the strain relaxation time and the compliance–frequency curves are illustrated in Figure 10.7.

10.7 Stress/strain relaxation and retardation

At constant stress (the constraint), the observed response is the change in strain and we call the phenomenon *strain relaxation* with *stress retardation*.

We have seen that if the stress on a sample of rubber is removed instantaneously, the strain falls to zero in an exponential way, i.e. constant stress equal to zero.

$$\gamma / \gamma_0 = \exp(-t / \tau)$$

where γ_0 is the initial strain under the applied stress and τ is the *strain* relaxation time. When the stress is periodic, working through the Laplace transform, the resulting periodic strain gives the complex ratio:

$$J^* / J_0 = 1 / \left(1 + \omega^2 \tau^2\right) - i\omega\tau / \left(1 + \omega^2 \tau^2\right)$$

In the same way, if the strain is raised instantaneously from zero to a given value and held constant, the observed response is the change in stress and we call it *stress relaxation* with *strain retardation*, i.e. under constant strain, the stress applied, equal to the internal force opposing strain, decays exponentially as the molecules take time to deform:

$$\sigma^* / \sigma_0 = \exp(-t / \tau)$$

where τ' is now the stress relaxation time. After taking the Laplace transform as before, we get the modulus–frequency relationship. The shape of this curve is just the mirror image (along the $\log(\omega\tau)$ axis) of the compliance curve. Since the logarithm of a reciprocal is minus the logarithm of the original quantity, a quick way of constructing the modulus curve is just to take the compliance curve and make $\log(\omega\tau)$ run from right to left instead of from left to right, at the same time moving the curve slightly along the frequency axis.

To understand the reason for the move, we have to appreciate that the time constant, τ, in the compliance equation is the strain relaxation time. However, for modulus we obtain the stress relaxation time τ'. Surprisingly these two relaxation times are not exactly equal.

Alternatively, we can take the reciprocal of the complex compliance, set the rubber modulus to zero, and change from τ to τ', when we get the result:

$$E^{*}/E_{0} = \omega^{2}\tau'^{2}/(1 + \omega^{2}\tau'^{2}) + i\omega\tau'/(1 + \omega^{2}\tau'^{2})$$

Note the $\omega^{2}\tau'^{2}$ term that appears in the numerator of the real part. You can see that this is necessary because without it the function would go from 1 to 0 instead of from 0 to 1. Also, the relationship between τ and τ' can be found by equating J_{0}/J^{*} to E^{*}/E_{0}.

Most periodic measurements apply an alternating stress and measure the alternating strain, giving the complex compliance according to the formulae we have discussed above.

Both the time dependent stress/strain measurements and the frequency dependent observations are collectively known as *dynamic mechanical relaxation*.

10.8   Relaxation and resonance

When we look at the curve of loss or energy absorption against frequency, it looks like a broadened version of the absorption curves that are familiar in spectroscopy. They, too, are energy loss curves, but are "resonance", not "relaxation". So what are the differences and the relationships between them? In fact they are two special cases of the same general response phenomenon. The spectrometers in a laboratory do not usually print out the real (energy storage) part of the complex response to the irradiating signal, but it is there none the less. Thus the whole response in spectroscopy is complex with real and imaginary parts, just as is the modulus or compliance in dynamic mechanical relaxation. Vibrational spectroscopy is particularly closely related, so we shall use a vibrational model to illustrate the connection.

Consider as a model a ball fixed to a support by a spring, and moving in a surrounding viscous medium (Figure 10.8).

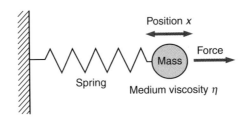

Figure 10.8 *Spring and ball in a viscous medium.*

When the ball is set in motion by a force, F, it oscillates under the Hooke's Law extension and compression of the spring, but its motion is retarded by the viscous drag of the surrounding medium. In infrared vibrational spectroscopy, the ball is an atom set in excited motion by the irradiation and there is no surrounding viscous medium. In dynamic mechanical relaxation, the ball is part of a molecule set in motion by an applied stress and other chains form the surrounding medium.

Then the total Newtonian equation of motion of the ball has the exciting force, F, the acceleration and deceleration of the vibration, the drag of the viscous medium and the Hooke's Law restoring force of the spring. This gives for the force and the restraints acting in opposite directions:

$$-F = a\left(\partial^2 x / \partial t^2\right) + \eta\left(\partial x / \partial t\right) + kx$$

where $a(\partial^2 x / \partial t^2)$ is the inertial acceleration term (resulting from the "push" of the force), $\eta(\partial x / \partial t)$ is the viscous drag of the surrounding environment and k is the Hooke's force constant of the spring. In an equilibrium situation the net force, F, is zero and

$$a\left(\partial^2 x / \partial t^2\right) + \eta\left(\partial x / \partial t\right) + kx = 0$$

Now, this is a second-order linear differential equation containing inertial, viscous and spring force terms. Unfortunately, such a differential equation does not have an explicit solution, so we have to look at limiting cases.

First, when the ball is moving slowly we note that this represents low vibration frequencies. Under these circumstances the inertial term, $a(\partial^2 x / \partial t^2)$, is small and can be ignored compared to the viscous term, $\eta(\partial x / \partial t)$, and the general equation reduces to:

$$-\partial x / \partial t = \left(k / \eta\right)x$$

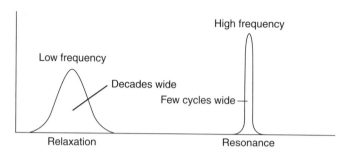

Figure 10.9 *Relaxation and resonance spectral peaks.*

This is the first-order linear differential relaxation equation. So we can say that at low frequencies the movement is relaxation.

On the other hand, at very high frequencies, the acceleration inertial term becomes large and more important than the first-order differential drag term; then:

$$-\partial^2 x / \partial t^2 = (k / a) x$$

This equation is soluble and the imaginary part of the solution is the resonance energy absorption.

The two extreme cases of absorption are shown in Figure 10.9.

The relaxation peak spreads over several decades of frequency as we saw from inserting numbers for $\omega\tau$ in the relaxation equation. But the sharp resonance peak only spreads over a few cycles in a very high frequency. So resonance is high resolution absorption but relaxation is very low resolution absorption. When the body is vibrating at 10^{12} Hz, as in vibrational spectroscopy, we have resonance energy absorption. However, if the body is vibrating at something like 10^0–10^3 Hz, we have relaxation energy absorption. Actually, most periodic stresses in dynamic mechanical phenomena are in the low frequency region and so the body exhibits relaxation. However, in certain ultrasonic and electromagnetic excitations, the frequency can be very high and the phenomena approach the resonance region where both inertial and viscous terms are important. Then the observed phenomena do not obey either the ideal resonance or the ideal relaxation equations, but something with some of the characteristics of both. There is then no explicit formula for the real and imaginary response characteristics as a function of $\omega\tau$.

10.9 Non-ideal dynamic mechanical relaxation behaviour

The above theory describes processes which conform to ideal relaxation behaviour and can be described accurately by a property decaying exponentially with time. The reduction in the population of an excited conformational state is one such process, and often follows closely the behaviour predicted above. Such processes are often referred to as *thermally activated*, the rate being controlled simply by the activation energy for the process. In polymeric systems, even simple conformational changes for groups in close proximity may possess activations energies which are slightly different from those of the isolated small molecules. This is a consequence of either variation in the stereochemistry of the attachment to the polymer backbone or to interaction with other groups brought into a surrounding position by the chain. So instead of there being one unique value of the relaxation time, there will be small variations from group to group. This variation can easily be accommodated in the above formulations by increasing the breadth of the relaxation process. This is done by introducing a distribution parameter α. The relaxation equation can now be written as:

$$R^*/R_0 = 1/(1 \pm \Sigma(i\omega\tau)) = 1/(1 \pm (i\omega\tau)^\alpha)$$

where α reflects the distribution of relaxation times associated with that process and has values ranging from 1 to 0.

In many materials, the structure of the polymer may give rise to more than one conformation change having similar activation energies. So, several different relaxation processes may occur close together in terms of their rates at a particular temperature. When the processes have slightly different activation energies, there will be a broadening of the relaxation behaviour. This situation can be accommodated by introducing a second distribution parameter β. The relaxation equation can now be written as:

$$R^*/R_0 = 1/\sum\left(1 \pm (i\omega\tau)\right) = 1/\left(1 \pm (i\omega\tau)\right)^\beta$$

where the parameter β has values between 1 and 0 and reflects the number of processes contributing to the overall relaxation feature. A general form of the relaxation equation which encompasses both possibilities has the form:

$$R^*/R_0 = 1/(1 \pm (i\omega\tau)^\alpha)^\beta$$

where the parameters α and β are really adjustable variables which allow the theoretical equations to be fitted to the experimental data. For simple situations, there is no ambiguity as to the changes involved. However, for polymers, if we attempt to ascribe specific molecular processes to the data then the situation becomes more complex. It is often necessary to supplement the experiments with other observations which allow the motion of specific chemical entities to be followed.

Unfortunately, the behaviour observed in dynamic mechanical relaxation rarely conforms to the simple ideal. Usually the data have to be fitted to equations that allow for the breadth to be increased, so accommodating a distribution of relaxation times and the possibility of more than one distinct process being involved.

One special relaxation process is worthy of note; this is the glass to rubber transition. Because the movement of the backbone segments depends on the availability of free volume, the temperature dependence is controlled not solely by the activation energy (as was shown in Chapter 4). As a consequence, the Arrhenius plots are curved and the processes are described by a relaxation equation where the parameter α is not unity.

Further reading

Havriliak S. and Havriliak S.J. *Dielectric and Mechnical Relaxation in Materials*, Hanser Publishers, Munich, 1997.

McCrum N.G., Reid B.E. and Wiliams G. *Anelastic and Dielectric Effects in Polymeric Solids*, John Wiley & Sons, London, 1967.

11

Acoustic (ultrasonic) relaxation

Polymers, and particularly rubbers, have very important vibration damping properties resulting from the energy loss when they are stressed and strained. A particular case of this is when the alternating stress is applied in the form of a sound wave. Then the property of technical significance is sound absorption. Furthermore, studies of acoustic relaxation shed further light on the dynamics and energetics of the molecular motions involved in the loss processes. In recent years, there has been significant interest in the use of high frequency ultrasonic radiation for the dispersion of particulate materials, and, in this context, it is important to understand the way in which high intensity ultrasound couples with polymers. Indeed, in certain situations ultrasonic irradiation can lead to degradation of the polymer chains.

In this section, we shall examine first the relaxation behaviour of a polymer material when irradiated with a sound wave, *acoustic relaxation*. Then we consider how the interactions may be influenced by increasing the intensity of the sound wave. Since most of the work in this area has been carried out in the ultrasonic frequency region, the phenomena are sometimes designated as *ultrasonic relaxation*. The irradiation of materials with high intensity ultrasonic waves is usually referred to as *sonochemistry*.

11.1 The sound wave

First, consider the frequency spectrum over which we can apply "sound" waves (Figure 11.1).

Audible sound constitutes a rather minor part of the whole spectrum. Infrasound defines those vibrations which are too low for human ears to hear, but which may be detected by earthquake monitoring equipment. Also, some animals appear to be sensitive to these long wavelength phenomena. The higher frequency ultrasonic waves are also not heard by humans, but they, too, may be detected by some creatures. The nomenclature split between ultrasonic and hypersonic waves relates to the limitations of highest frequency wave propagation in air. So the overall sound wave spectrum covers a wide frequency range and thereby allows coupling with a wide range of molecular processes.

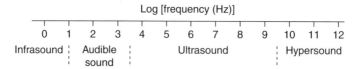

Figure 11.1 *The sound frequency spectrum.*

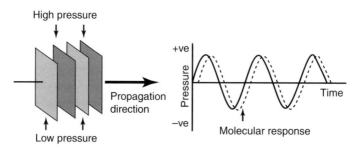

Figure 11.2 *The sound wave.*

A sound wave can be considered as a propagating one-dimensional pressure fluctuation which, viewed from a stationary point in space, appears as a periodic compression and rarefaction of the transmission medium. In other words, the wave front can be visualised as a pressure plane in two dimensions perpendicular to the direction of propagation (Figure 11.2).

Now, a one-dimensional pressure wave can be resolved into three-dimensional pressure plus/minus two-dimensional shear (Figure 11.3).

For an accurate analysis of the propagation, one must include the shear component. However, for many simple situations the relaxation processes associated with shear occur outside the frequency range of interest, and so for the rest of this section we shall ignore the shear component, and consider only the response of a system to a three-dimensional pressure wave.

For most systems the change from high to low pressure in a sound wave occurs in such short times that the system cannot exchange heat with its surroundings, and so the pressure change is *adiabatic*. However, adiabatic compression causes the molecules to undergo more collisions and so raises the temperature. Adiabatic rarefaction lowers the temperature. So, the sound wave is effectively

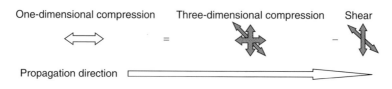

Figure 11.3 *Resolved pressure vector components of the sound wave.*

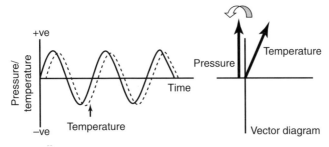

Figure 11.4 *Pressure and temperature waves.*

a temperature wave. When the temperature rises, molecules take energy in. When the temperature falls, molecules give energy out. However, the molecules need time to take energy in and put it out, so the temperature wave is moderated by the molecules and its phase lags behind the phase of the pressure wave. This phase lag is a measure of the time needed for the energy intake and output. In other words, the moderated temperature wave is the lagging response to the pressure constraint (Figure 11.4).

11.2   Measured sound parameters

When a sound wave of frequency f in hertz propagates through a sample, the propagation occurs with a velocity, v, giving the sound wavelength, λ. If a pulse of sound is sent from a transducer through a sample of known thickness, and received on a second transducer as it leaves the sample, the time difference as the pulse enters and leave the sample can be measured, so giving the velocity, v. Also the amplitude of the pulse can be measured as it enters and leaves the sample, giving the absorption coefficient, α. These measurements allow us to characterise all the transmission properties of the sample. It is usual to portray the data in the form of three parameters: v, the velocity, α/f^2 as a measure of the absorption, and another such measure, $\mu = \alpha\lambda$.

While the velocity, v, is the familiar real part of the transmission parameters, α/f^2 and $\alpha\lambda$ represent rather different portrayals of the imaginary component. However, just as in typical relaxation curves against the logarithm of the frequency, these parameters exhibit relaxation shapes as the frequency changes from being lower than the rate of the molecular energy in–out process, to being higher than the rate of the molecular response (Figure 11.5).

These curves contain essentially two different types of information. The x-axis, being frequency, contains kinetic information about the rate of the molecular process involved. The y-axis, being the amplitude of an energy effect, contains thermodynamic information about the energies involved in the molecular response process.

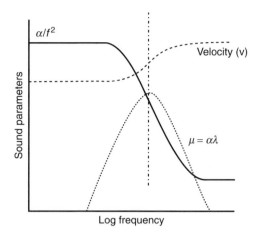

Figure 11.5 *Acoustic relaxation.*

11.3 Two-state energy model

The interaction of a molecular system with a sound wave is usually interpreted on the basis of a two-state energy model. When the temperature rises, molecules in the lower energy state absorb energy and are raised to the higher state. Conversely, when the temperature drops, molecules in the higher state give out energy and drop to the lower state. The response of the system to the changing temperature constraint is thereby a change in the relative populations of the two states. The relaxation frequency is then the frequency of interchange between the states. There may well be a rate-determining activation energy barrier to the state interchange process, so we can enter it, too, on an energy level diagram (Figure 11.6).

The thermodynamic amplitude information, measuring the amount of energy moving in and out of the molecules, can be used to evaluate ΔE^0. The kinetic frequency information, if studied as a function of temperature, can be put into the Arrhenius equation to yield the value of the activation energy, $\Delta E^{\ddagger}$.

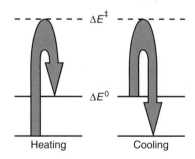

Figure 11.6 *Two-state energy interchange.*

For a relaxation to be observed, the frequency of the interchange process must be similar to the sound frequency, and, in addition, the energies involved in the process must be comparable to the energies available in the temperature variation of the sound wave. Both of these criteria can be met by some of the conformation changes we have been discussing earlier.

11.4 Polymer internal rotation

We have seen that internal rotation of a unit in the polymer chain is an important process for dynamic mechanical relaxation, causing the glass to rubber transition. In mechanical studies, we were able to observe the frequency of this process, but were not able to evaluate the energy differences between the important rotational states. It was possible to draw only qualitative conclusions based on the expected differences in steric interaction or electrostatic forces. However, with acoustic relaxation, we have a set of observations in which the amplitudes directly yield the energy differences of interest. For two polymers, respectively of high and low energies of rotation, the absorption data yield a comparison of both the energy difference between rotational states (from the amplitudes) and the activation energy for rotation (from the rates) (Figure 11.7).

These conformational energies, ΔE^0 and $\Delta E^{\ddagger}$, have been evaluated for a number of polymer chains and related small molecules. An interesting finding is that the energy difference between the boat and chair forms of cyclohexyl side groups in polymer chains is almost the same as that of cyclohexane in liquid form, confirming what was met already when discussing low temperature transitions in glasses!

The last remaining piece of the jigsaw is to determine how these conformation energies determine the way that the mechanical loss behaves both with

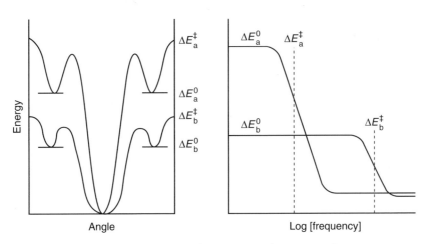

Figure 11.7 *Rotational energies and acoustic relaxation.*

temperature and with steric hindrance. To do this, we remember that acoustic vibration is just high frequency mechanical vibration, and then dynamic parameters so determined are simply the high frequency mechanical figures. Now, the energy absorption amplitude, which is the dynamic mechanical loss, is related to ΔE^0 and thermal energy, RT, by the *Schottky equation*.

First, we examine how much energy can be carried from the lower to the upper state as a function of ΔE^0. So energy absorbed in changing from the lower to the upper state is zero when the energy difference is negligible, rises to a maximum when the energy difference is finite but not too large compared to RT, and then falls back to zero again when the energy difference is too large for excitation (Figure 11.8).

It is possible to treat this with the thermodynamic equations relevant to the absorption of energy through such excitation, and the result is the Schottky curve relating loss to energy difference (Figure 11.9). The Schottky equation allows separation of the enthalpy ΔH^0 and the entropy ΔS^0, the magnitude of the energy loss being a function of both parameters.

The actual values depend on the entropy change between the states, a factor we have conveniently ignored in this oversimplified introduction to the loss process. Inclusion of the entropy term into the free energy produces a set of curves which are slightly displaced from one another, but have essentially the same shape.

The origin of the entropy change is easy to visualise in the case of a simple conformational change. The higher energy form has greater entropy having more than one possible conformation, whereas the lower energy form is a single conformation. The change from the higher energy to the lower and vice versa will involve not only an enthalpy, but also an entropy, change. The slopes of the graphs are very similar on the higher temperature side of the peak but are displaced along the $\Delta H^0/RT$ axis as a consequence of the entropy change. In many polymeric materials, the energy profile associated with the relaxation, although

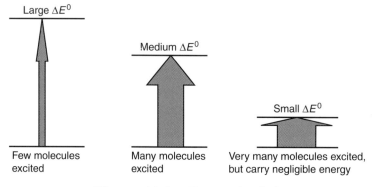

Figure 11.8 *Energy absorbed.*

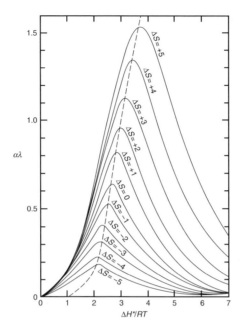

Figure 11.9 *Schottky curve of loss against* $\Delta E^0/RT$.

involving the movement of a number of segments in a cooperative manner, may still accurately be approximated by the simple profiles presented in Figure 11.7.

There are several points to note about the Schottky curves. Firstly, the energy absorption at a given temperature first rises then falls with increasing ΔH^0. Secondly, the energy absorption for a given two–state system first rises then falls with increasing temperature. In fact, these observations are the quantitative representation of the loss changing with steric hindrance and temperature at the transition temperature, a fact that was derived by qualitative arguments earlier. So, at long last, we have a sound theoretical backing for the energy loss behaviour of various polymers at various rates and temperatures.

11.5 High intensity ultrasonic waves in liquids

It is relatively common for ultrasound to be used to aid dispersion of materials in a liquid. In recent years, there has been much interest in the possibility of inducing chemical reactions as a consequence of the sound wave propagating through the liquid, and so the subject of *sonochemistry* has emerged. In the literature, much attention has been focussed on what happens when *cavitation* is induced in a liquid. If the amplitude of the pressure wave is increased to a very high level, the adiabatic heating and cooling will result in the liquid being stressed to such an extent that local heating occurs and bubbles of vapour can be formed. The collapse of these bubbles can give rise to very high temperatures and many of the effects observed are attributed to an increase in the rate of the reaction caused by these. However, if we expose a polymer solution to high intensity ultrasonic waves, controlled to be just below the point at which cavitation occurs, degradation of the polymer can take place. If the polymer molecules were to be degraded simply as a consequence of radicals produced by the high temperatures then we would expect that the polymer chains would be chopped up in a random fashion. In practice, though, the chains are reduced in size by approximately half their original length. In Chapter 3, the concept of normal mode motions was introduced. The lowest frequency movement corresponds to the first normal mode, in which the motion corresponds to a deformation with the chain having a node at the centre. The node will correspond to the point at which the chain is being stretched by the motion of the rest of the chain and scission occurs at this point. Careful examination of the molar mass distribution of the polymer molecules after sonic irradiation indicates that the shift in the molar mass distribution is consistent with chopping the chain in two. A further important observation is that if the first normal mode occurs at a frequency which is higher than that used to produce the ultrasonic wave then the coupling with the normal mode motion does not occur and degradation is not observed.

11.6   Acoustic modes in solids

In crystalline solid polymers, not only do the chains vibrate in the normal manner which we associate with infrared or Raman modes, but it is also possible for the all-*trans* section of a chain such as polyethylene to undergo an accordion type of motion. The element which is involved in the motion is constrained by the size of the crystal lamellae described in Chapter 6. The motion is not a relaxation and it gives rise to a resonance observed in the far infrared or Raman spectrum. Typically, crystalline polyethylene will possess a vibration band at about $120\,cm^{-1}$ which is associated with this accordion motion. The precise position depends on the length of the lamellae, the shorter lamellae having a higher frequency of resonance. This collective vibration of the chains is quantised, and is a *phonon*. These well defined acoustic vibrations are very important in understanding the temperature dependent dynamic behaviour of crystalline solid polymers.

Further reading

Pethrick, R.A. *Acoustic Properties*, in G. Allan, J.C. Bevington (Eds.), *Comprehensive Polymer Science, Volume 2 Polymer Properties*, Pergamon Press, Oxford, 1989.

12

Dielectric relaxation

12.1 Band structure of insulators, semiconductors and metals

Solid materials can be divided into three classes: insulators, which do not readily conduct electricity; metals, which easily conduct electricity; and semiconductors, which in the presence of an external stimulus such a light or a voltage will conduct electricity. Whether or not a material conducts electricity depends on the energy structure of its highest energy electrons.

When isolated single particles come together into a many-particle system, the highest energy electrons interact. As a result, what were single energy levels in the isolated particles become bands of energy levels in the solid. Chemists are familiar with the way single levels in an isolated atom are split into two (bonding and antibonding) when the atoms unite to form a diatomic molecule. The formation of bands in a many-particle solid is just this splitting reproduced many-fold (Figure 12.1).

If any electrons are in the upper band, they are mobile (spread) over the interacting particles. Thus they can carry electricity across the particles and the energy band is called the *conduction band*. On the other hand, electrons in the lower band are immobilised on the atoms or in the valence bonds of the particles, and so the band is called the *valence band*. Although such electrons cannot conduct electricity, if there is an electron missing from one of the bonds, this is called a positively charged *hole*, and a hole can "hop" from bond to bond, thus causing the movement of positive charge. In terms of molecular systems, the valence band is referred to as the Highest Occupied Molecular Orbital, HOMO, and the conduction band is the Lowest Unoccupied Molecular Orbital, LUMO. The conduction process is therefore the HOMO to LUMO transition and this is important in intrinsically conducting polymer systems.

Then solids are classified in terms of the energy separation of, and electron distribution between, the valence and conduction bands (Figure 12.2).

In an insulator, the forbidden energy gap between the valence and conduction bands is so large that the thermal energy, kT, per electron is too small to elevate electrons to the upper level, and the solid is an insulator at all temperatures.

In a semiconductor, the energy gap is less than the thermal energy, kT, per electron, which can elevate electrons to the upper conduction levels. Thus

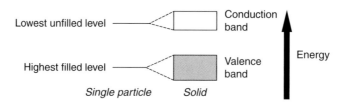

Figure 12.1 *Molecular electron energy levels split in a solid.*

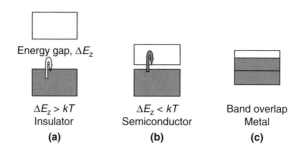

Figure 12.2 *Band structure: (a) insulators, (b) semiconductors, and (c) metal-type conductors.*

conduction is possible, and it increases because raising the temperature elevates more electrons to the mobile levels. Excitation from the valence to conduction bands in semiconductors can also be achieved if a photon is absorbed by the molecule in the ground state. This leads to the excitation of an electron and so observation of photoconduction. It is also possible to input energy by application of a voltage bias to the system, also allowing an electron to be pulled up to the conduction band. This latter process forms the basis for many transistor and computer devices.

In a metal, the valence and conduction bands overlap (in the ideal case the upper band is half filled). Thus a metal is a conductor, regardless of the magnitude of the thermal energy. In a metal, the electron is not bound to any particular atom so is able to diffuse around the lattice.

Most polymers are insulators, and it is with these that we shall be concerned. However, it must be noted that certain polymers with conjugated backbones are semiconductors, and these are rapidly becoming of increasing importance in a variety of uses. For example, polymer semiconductors are used in spacecraft batteries, in transistors and in electro–luminescent display devices. Nevertheless, the second most important use of polymers, after uses depending on their mechanical properties, is as electrical insulators. Almost all electrical wiring is insulated by a polymer covering; the electronic base board and other components are separated electrically from each other by polymers, while devices such as capacitors use polymers as the capacitive charge storage material.

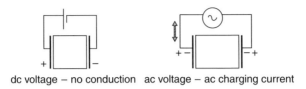

Figure 12.3 *Charging and discharging a capacitor.*

Now, when an insulator is placed between two conducting electrodes and subjected to a steady (dc) voltage, it prevents conduction between the electrodes. However, when it is subjected to an alternating (ac) voltage, it functions as a capacitor, and there is a charging current as the capacitor charges and discharges under the influence of the changing field. As we shall see later, the magnitude of the charge which can be stored in a capacitor will depend upon the permittivity of the polymer which fills the gap between the metal electrodes (Figure 12.3).

12.2   Complex permittivity

Before looking at the effects of frequency, let us revise the spread of the electromagnetic spectrum (Figure 12.4).

The voltage signal is propagated by different methods at different frequencies. Up to the megahertz region, ordinary wires are adequate. At high megahertz to gigahertz frequencies, it is necessary to confine the signal between the inner and outer electrodes of coaxial cable. At gigahertz frequencies, the signal propagates down a rectangular waveguide. Finally, infrared waves are transmitted by radiation through free space. The higher the frequency is, the shorter the wavelength. In the gigahertz frequency range, the wavelength is of the order of a few centimetres and becomes shorter as the frequency is increased. At low frequency, the wavelength is such that the potential at all points on the electrodes surrounding a sample will be the same, whereas at megahertz and gigahertz frequencies, it is possible to see the signal moving across an electrode structure and an analysis of this propagation behaviour is required to calculate the dielectric properties of a material.

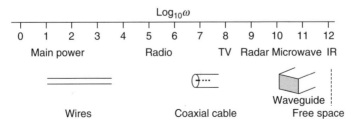

Figure 12.4 *The electromagnetic spectrum.*

The response of an insulating material to an alternating voltage is called its *permittivity*. It is a measure of the way the charge carriers move under the influence of the electric field. As such, it is the electrical analogue of compliance. Compliance measures the ease with which molecules move under an applied mechanical force, while permittivity measures the ease with which charge moves under the influence of an applied voltage force. As a result, all the equations that are used for mechanical relaxation should apply equally to dielectric relaxation, simply by replacing stress and strain with the appropriate electrical parameters.

So, to move under the influence of an applied alternating voltage, what do charge carriers need? They need energy and time! Consequently, the carrier movement and the resulting charging current lag behind what would be expected from the phase of the voltage. The electrical case, though, resembles more closely what has been discussed under viscosity than under elastic modulus. This is because the charging current is proportional to the differential of the voltage change (dV/dt), not the voltage itself. So the phase of the alternating charging current should lead the phase of the voltage by 90°. Then the complex response lags behind the ideal 90°. Again, as in mechanical relaxation, we can see this when the charging current is shown as rotating vectors in phase space in an Argand diagram (Figure 12.5).

The complex charging current vector can be resolved into components 90° ahead of, and in phase with, the voltage. The ideal component, 90° ahead of the voltage, is real, while the component in phase with voltage represents an ohmic current, energy lost, and so is imaginary. The imaginary component is responsible for the heating effect which results from the polarisation of the system.

Just like compliance, the permittivity is complex and can be resolved into real and imaginary components:

$$\varepsilon^* = \varepsilon' - i\varepsilon''$$

where ε is the permittivity. The real permittivity, representing energy stored and recovered, is measured by capacitance, while the imaginary permittivity,

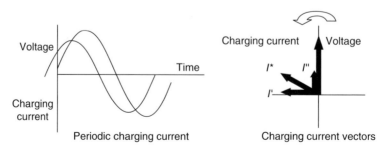

Figure 12.5 *Complex charging current as vectors in the Argand diagram.*

representing energy lost by ohmic conduction, is measured by conductance. A variety of different methods have been developed to allow direct measurement of the capacitance and conductivity of materials. Computer driven measuring equipment effectively can allow measurement to be made over the frequency range 10^{-5} Hz to ~ 100 GHz. Of course, the techniques used differ according to the frequency range of interest.

12.3 Time and frequency dependences

We are now in a position to use our relaxation equations to express the time and frequency dependences of the capacitive charging and so of the complex permittivity.

Starting with the charge as a function of time:

$$Q/Q_0 = \exp(-t/\tau)$$

where τ is the time constant for an equivalent resistor–capacitor (RC) circuit, or the relaxation time of the polymeric capacitor. Then, once again, going to the Laplace transform of this time function, and going from charging current to permittivity:

$$\varepsilon^*/\varepsilon_0 = 1/(1 + i\omega\tau)$$

where again we note the positive sign. Rationalising by multiplying numerator and denominator by the complex conjugate, this time $(1 - i\omega\tau)$, gives:

$$\varepsilon^*/\varepsilon_0 = 1/(1 + \omega^2\tau^2) - i\omega\tau/(1 + \omega^2\tau^2)$$

$$= \varepsilon' - i\varepsilon''$$

This is shown in Figure 12.6.

The capacitive (real) and conductive (imaginary) components of the complex permittivity are exactly the same shape as those for the complex compliance that were presented in Chapter 10.

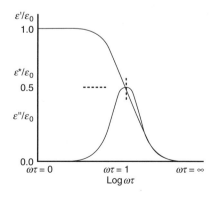

Figure 12.6 *Real and imaginary permittivity as a function of log frequency.*

12.4   Dipole orientation polarisation

The essential feature of dielectric capacitance is that electric charge moves under the influence of the applied voltage to charge the capacitor, but then is trapped and cannot move further through the capacitor and through the contact electrodes. This is called *polarisation* of the sample. So, what are the molecules doing? Somehow the molecules are participating in the polarisation process, charged entities within them being attracted towards the oppositely charged electrode, but being unable to flow as in dc conductivity. While a molecular dipole will be orientated by the applied field, the charge movement is restricted to polar bonds. The summation of the effect of the orientated dipoles gives rise to the overall sample polarisation. In some systems, free ions may exists and partial movement followed by trapping also can give rise to polarisation and permittivity.

The charge movement that is most similar to the polymer chain movements discussed in Chapter 1 is dipole orientation. Under the influence of an electric field, dipoles that are originally randomly orientated are turned so that the charged ends are attracted to the appropriately charged electrodes (Figure 12.7).

This diagram exaggerates the extent of orientation, which in fact is only fractional. Nevertheless, the resultant charge distribution over the whole sample is such that the sample interface next to the negative electrode carries a net positive charge, and the interface next to the positive electrode carries a negative charge. This is the sample polarisation or capacitance. It results in a high permittivity.

We note two important characteristics of this orientation:

(1)   Because the positive and negative ends of the dipole are connected by a covalent bond (or sometimes bonds), the charges cannot separate to conduct electricity.
(2)   The dipoles need time to rotate.

If the frequency of the alternating voltage is too high, there is not enough time for the dipoles to rotate, there is no polarisation, and consequently the permittivity is low. So the permittivity falls off with increasing frequency just as did the compliance.

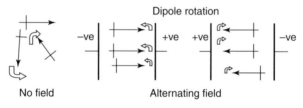

Figure 12.7   *Dipole orientation polarisation.*

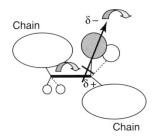

Figure 12.8 *Dipole orientation by internal chain rotation.*

Indeed, if the dipole is rigidly attached to the polymer chain backbone, the dipole orientation takes place by exactly the same internal rotation as led to the strain under an applied stress (Figure 12.8).

The dimensionless $\varepsilon^{*}/\varepsilon_{0}$ against log frequency curves superimpose on those of J^{*}/J_{0}. So, too, do the curves against temperature.

It will have been noticed in the above that it is the dimensionless normalised curves that are identical. Of course, the amplitude factors ε_{0} and J_{0} have different units and may be very different in magnitude. Thus a transition that is very large in compliance may be very small in dielectric permittivity (small dipole moments involved). Conversely, a transition that is difficult to spot in compliance measurements may have large amplitude in the permittivity relaxation (large dipole moments orientating). So dielectric relaxation can be used to measure the polymer transitions in exactly the same way as can dynamic mechanical relaxation.

A significant aspect of the change in permittivity is seen because the real permittivity, ε', is in fact the dielectric constant. So a polymer will have a high dielectric constant if the chain has high dipole moments and is free to undergo internal rotation. However, when the rotation cannot take place, the dielectric constant is much lower. So a polymer glass will always have a lower dielectric constant than a polymer rubber of the same molecular polarity.

Of equal importance is the fact that a polymer will show electrical loss, as ac conduction, in the transition region. So when polymers are used as ac insulators, it must always be well away from the glass to rubber, or even the β, transition (Figure 12.9).

Since rotation of a dipole rigidly attached to a polymer chain is much more restricted than rotation of the same dipole in a small molecule in a liquid, the relaxation or transition frequency is much lower than in the small molecule case. Thus dipoles in the rubber state orientate in the kilohertz to megahertz frequency region, while those in small molecule liquids rotate in the gigahertz to terahertz region. Consequently, when polymers are used for electrical insulation, particularly at radio frequencies, care has to be taken to ensure that the polymer is either in the high permittivity low frequency (rubber) region, or in the low permittivity high

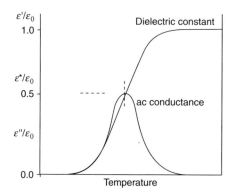

Figure 12.9 *Complex permittivity, dielectric constant and conductance as a function of temperature.*

frequency (glass) region. The dielectric constant, too, is often lower than that of the analogous small molecules of identical polarity, especially at higher frequencies.

As was presented in Chapter 4, for the working range of a tough glass, the temperature range between transitions can also define the *capacitive working range* of the polymer, at least for applications where precision and consistency are required. As we saw when examining the modulus of polycarbonate glasses, this working range is a function of the frequency of the applied constraint, here voltage, becoming reduced at the highest frequencies (Figure 12.10).

This concept of the working range applies to polymers between the glass to rubber and melt transitions just the same as between the α (glass to rubber) and β transitions. In practice, for high quality capacitors, it is essential to ensure that there are no mobile charges in the polymer. During the synthesis process, ionic catalysts are often used and residues of these can enable ionic conduction. Removal of these residues ensures that the measured frequency response arises purely from movement of the molecular dipoles.

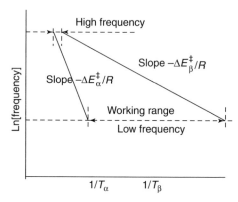

Figure 12.10 *Arrhenius diagram of α and β transitions.*

12.5 Interfacial polarisation

We come now to a quite different form of polarisation, although once again charge is trapped at the extremities of dipolar entities. This is the electrical effect when a discontinuous conducting phase is dispersed in a continuous insulating phase. The phenomenon was studied first by Maxwell, and then by Wagner, and most recently by Sillars. So, for polymer systems, it is called Maxwell–Wagner–Sillars polarisation.

Consider a rectangular particle, able to transport charge carriers, surrounded by an insulating medium, into which the charge carriers cannot move. Then, under the influence of an electric field, the charge carriers move through the conductor until they are trapped at the interface with the insulator.

The first thing to note is that if the conducting phase has many charge carriers, and is of macroscopic dimensions, the dipolar charge and distance of separation are very large, leading to colossal dipole moments, huge polarisations and so to enormous permittivities and losses (Figure 12.11).

The second point is that the charge carriers need time to move through the conducting phase, a time which will depend on the mobility of the carriers in the conducting medium. If this mobility is low then a relaxation will be observed at low frequencies. On the other hand, if the mobility is high then the relaxation will be observed at high frequencies.

Now, the conductance of the dispersed phase depends on both the number of charge carriers and their mobility, so as a result we can get very large relaxation amplitudes occurring at widely differing frequencies. We shall take two extreme examples, both of considerable technical significance.

12.6 Carbon fibre composites – stealth aircraft

Carbon fibre reinforced resins are widely used where great strength, flexibility and lightness are required, as in the aerospace industry, as well as in automobiles, golf clubs, tennis rackets, and so on. Now, carbon fibres are predominantly graphite, which is a good conductor of electricity (although less so than metals). Charge is carried by electrons in the conduction band and also by holes in the valence band. The carriers are large in number, but have a mobility about one thousandth that of electrons in a good metal conductor. In fact, the mobility is

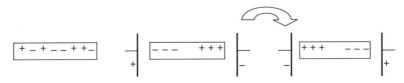

Figure 12.11 *Charge carrier movement to the limits of the conducting phase.*

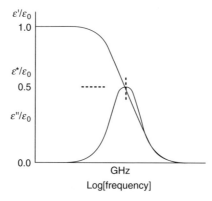

Figure 12.12 *Complex permittivity showing the absorption of radar electromagnetic radiation.*

such that it takes a fraction of a nanosecond for the carriers to move from one end of a carbon fibre to another. However, once the carriers reach the end of the graphite, they are trapped and cannot move further in the insulating polymer resin. The result of the high number and the high mobility means that there is a large polarisation which relaxes in fractions of a nanosecond, or which has a relaxation transition at gigahertz frequencies. Below the relaxation frequency, the real permittivity is very large, but of greater technical significance is the fact that around the relaxation frequency the imaginary permittivity, the energy absorption, is also very large. In other words, around the relaxation frequency, the composite can absorb large amounts of electrical energy. This is true whether the energy is supplied through wires and electrodes, or as electromagnetic radiation.

Electromagnetic radiation at gigahertz frequencies is used in radar detection of aircraft. Therefore aircraft with structural members and skin coatings of carbon fibre will absorb much of the energy of incident electromagnetic radiation (radar) and not reflect it. In fact, if the fibres are aligned with the electric field vector, absorptions thousands of times greater than those of unfilled polymers can be obtained. Combined with the angular geometry of the aircraft surfaces, this means that the detector linked to the radar signal generator receives no reflected waves. This is the basis of "invisible" or "stealth" military aircraft. Now many radar absorbing coatings can be obtained using other materials as the dispersed phase (Figure 12.12).

12.7 Electronic baseboards

Electronic baseboards are used for mounting the chips, transistors and other elements of electronic circuitry. They are generally made of either phenolic resin reinforced with cellulose (paper), or epoxy resin reinforced with glass fibre. Their principal purpose is to give physical support to the circuit elements while insulating them from

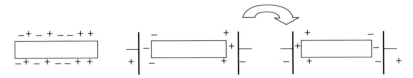

Figure 12.13   *Charge carrier movement along the interface.*

each other. It is important, therefore, that the board be both a dc insulator and also an ac insulator at all the frequencies likely to be encountered in use.

In early boards, the surfaces of the cellulose in particular, but also of the glass, were hydrophilic, and in warm humid conditions water became absorbed at the resin–filler interface. This absorbed water could either self-ionise into hydrogen and hydroxyl ions, or facilitate the ionisation of impurities at the interface. As a result, charge carriers could be formed at the interface (Figure 12.13).

The movement of ions along the surface of the dispersed cellulose or glass is a relatively slow process, and requires times around the order of seconds. As a result, a dielectric relaxation arises in the hertz frequency region. The amplitude of this relaxation depends on the number of ions formed, and so on the amount of water absorbed. This, in turn, depends on the relative humidity of the sur-rounding atmosphere.

At frequencies around the hertz region, the boards with absorbed water show ac conductivity, so lose their insulating powers with consequent failure of those electronic processes at these frequencies. Unless steps are taken to inhibit mois-ture uptake, the magnitude of the ac conductance is greatest in a humid atmos-phere (Figure 12.14).

Again, permittivities and losses a thousand times greater than those of unfilled polymers may be encountered.

Although most ac electronic circuits function at frequencies much higher than these loss frequencies, failures have been experienced when high fre-quency signals are pulsed at these very low frequencies, since the pulse repetition frequencies may fall in the loss region.

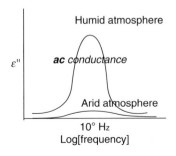

Figure 12.14   *Loss conductivity in desert (arid) and tropical (humid) atmospheres.*

The absorption of water is greater in the cheaper cellulose-filled phenolic boards than in the more expensive glass-filled epoxy boards. Nevertheless, modern boards of both types are now fabricated with surface treatment of the filler to minimise water absorption.

12.8   Collisional polarisation

Perhaps the highest frequency of dielectric relaxation, verging on resonance, occurs in the sub-millimetre microwave to far infrared region of the electromagnetic spectrum. At these frequencies phenomena are found that have the characteristics of relaxation at the lower frequencies, but change to the characteristics of resonance at the higher frequencies. Here polarisations occur, or fluctuate, in picoseconds, so giving rise to dielectric phenomena in the high gigahertz to terahertz region of the spectrum.

One phenomenon of interest, and also technical importance, is the electronic polarisation that occurs when two molecules, or groups on molecules, undergo collisions. In solids and liquids. molecules are in a constant state of motion, colliding with, and rebounding from, their neighbours. Of course, in polymers, this is done by sections of the chain rather than by the whole chain acting as a single integrated entity. When two molecules collide, they rebound because the diminishing separation leads to increasing repulsion energy. This in turn has its origin in the fact that the approach tries to force the electrons of one molecule into the space occupied by the electrons of the other. In order to avoid violating the Pauli exclusion principle, the electrons are forced into higher energies and directed away from the area of impact. Thus each molecule, at the moment of impact, becomes polarised with the negative end away from the contact interface (Figure 12.15).

When the frequency of an applied electric field is the same as the collision frequency, the field and the fluctuating dipoles interact and we get relaxation or resonance phenomena. In other words, the frequency of maximum loss, of energy absorption, is the molecular collision frequency.

Now, all molecules exhibit this electronic polarisability, the effect being greatest when there are most electrons least strongly bound to the molecular nuclei. For polymer units this means that:

$$-CH_2-CH_2- < -CH_2-CHCH_3- < -CH_2-CH(C_6H_5)-$$

Figure 12.15   *Fluctuating collision dipole.*

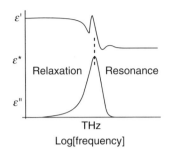

Figure 12.16 *Complex permittivity due to collision polarisation.*

Polymers with multi-electron atoms, like halogens, have even higher polarisations. Aromatic rings can become distorted on collision and this will also result in a change in the charge distribution. So this, too, results in collisional polarisation.

In our permittivity units, the complex response of the system looks like both relaxation and resonance (Figure 12.16).

When the absorption is displayed in the spectroscopic convention of an absorption coefficient, the relaxation component appears as a curve rising smoothly with increasing frequency.

If sub-millimetre (high gigahertz) microwaves (as in telecommunication systems) are going to pass through, or impinge on, polymer materials (as in filled wave guides) then the lowest power absorption will be shown by polyethylene, with all other polymers showing greater polarisability and so greater absorption.

Further reading

Kremer F. and Schonhals A. (Eds.) *Broadband Dielectric Spectroscopy*, Springer, Heidelberg, 2003.

Ku C.C. and Liepins R. *Electrical Properties of Polymers*, Hanser Publishers, Munich, 1987.

13

Photophysics of polymers (excited state relaxation)

Readers interested in polymer science should be familiar with the photochemistry of polymers. Photochemistry plays a role in polymerisation reactions, in the degradation of the backbone chain, or in cross-linking of different chains, all of which can be initiated by light Further, photochemistry has been used in a variety of different ways to change and probe polymer structure and intermolecular interactions. However, there are a number of primary physical processes, which take place between the photon impinging on the polymer and the commencement of chemical reaction. It is these primary physical processes that are the subject of this chapter.

13.1 Single molecule processes and relaxation

The very first processes that occur when a photon meets a single molecule can be summarised in an energy level diagram called the *Jablonski diagram* (Figure 13.1).

The important energy levels of the molecule are the electronic ground state (usually a singlet state with all electrons having paired spin) and the first excited electronic singlet state, in which the excited electron retains the same spin state. Then there is also the possibility of a triplet state having two electrons of unpaired spin. This is lower in energy than the excited singlet state but higher than the singlet ground state. Each of these electronic levels has its own set of vibrational levels.

The five important energy processes are:

(1) Absorption of the incident photon with an increase of energy from the lowest vibrational level of the ground state to an upper vibrational level of the excited state. The promotion of the electron from the ground to the excited state is very fast compared to the vibrations in the molecule, a fact that is called the Frank–Condon principle.

(2) Radiationless loss of vibrational energy as heat when the molecule relaxes to the ground vibrational state of the excited electronic state.

(3) Radiationless transfer to the triplet state. This change of spin of the electron does involve interaction between the electron and some vibration of the molecule.

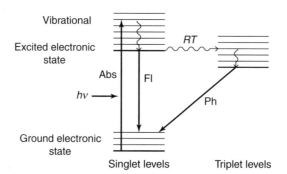

Figure 13.1 *Jablonski diagram for a single chromophore. Abs is absorption, Fl is fluorescence emission, Ph is phosphorescence emission and RT is radiationless transition.*

(4) Emission of *fluorescence* as the molecule relaxes from the excited singlet state to upper vibrational levels of the ground singlet state.

(5) Emission of *phosphorescence* as the molecule relaxes from the triplet state to the ground singlet state. (The word *luminescence* is used to cover both fluorescence and phosphorescence.)

The Jablonski diagram shows that the energies involved are largest for absorption, then less for fluorescence, and least for phosphorescence. As a result, the radiation frequencies involved fall in the same order, as can be seen in spectra of amplitude (emission or absorption) against wavelength (Figure 13.2).

Of particular importance are the times which the molecule spends in the various states. Absorption occurs in around 10^{-13} s. This can be considered to be instantaneous and is roughly a factor of ten times faster than the molecular vibrations. The time in the singlet excited state is of the order of nanoseconds, while the time in the triplet sate can be anything from microseconds, through milliseconds, to

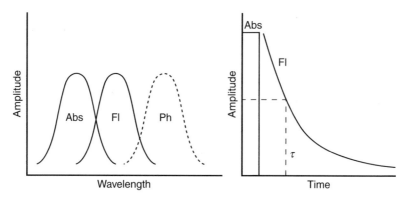

Figure 13.2 *Absorption, fluorescence and phosphorescence spectra and fluorescence lifetimes.*

minutes. The longer lifetime for the triplet state is because the triplet to singlet transition is a forbidden electron process. The molecule may stay in the triplet state until deactivated by some mechanism other than radiation of phosphorescence.

Since the emission intensity is proportional to the number of molecules in the emitting state, we can observe the time dependence of the fluorescence or phosphorescence and thereby study fluorescence (or phosphorescence) relaxation. If $n*$ is the number of molecules in the excited state:

$$n* = n*_0 \exp(-t/\tau)$$

$$I_{Fl} = I_{Fl,0} \exp(-t/\tau)$$

where I_{Fl} is the intensity of fluorescence and τ is the lifetime of the excited state (its relaxation time). To measure this most accurately, it is desirable to use a pico- or femto-second pulsed laser for the excitation, and sub-nanosecond single photon counting for the luminescence detection. This lifetime is a sensitive measure of the other things that can happen to the energy in multimolecular systems.

13.2   Two-molecule systems: exciplexes and excimers

Now consider two molecules, A and B, in such proximity that the electrons of one can interact with those of the other. For simplicity we shall confine ourselves to singlet energy levels, although many of the concepts that we shall explore are also relevant to triplet states.

The various energy processes can be summarised on an extended Jablonski diagram (Figure 13.3).

Two new phenomena can now occur. If the incident photon is absorbed by molecule A, it may be possible for the energy to undergo a resonance transfer to molecule B.

$$A* + B \rightarrow A + B*$$

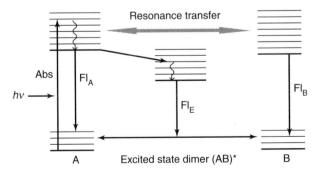

Figure 13.3   *Two-molecule energy level diagram.*

The super script * designates the electronic excited state of the molecule. This process will be attended by a diminution of the fluorescence from molecule A, and the appearance of fluorescence from molecule B.

It may also be possible for the two molecules to form an excited state dimer.

$$A* + B \rightarrow (AB)*$$

This dimer may have a lower energy than the excited states of either of the individual molecules. However, it will require the two molecules to be in very close proximity (the dimer bond distance), which will be less than the normal Van der Waals separation of the ground state molecules. This separation will, then, be high energy and repulsive in the ground state. As a result, the dimer will have a stable existence only in the excited state. The excited state dimer is called an *exciplex* when A and B are different molecules, and an *excimer* when two identical molecules (A + A) are involved.

So now three different peaks of fluorescence may be observed, each with a different lifetime. Indeed, the resonance transfer time and the dimer formation time are often comparable to the fluorescence lifetime, so both the build up and the decay of these other excited state fluorescences can be observed (Figure 13.4).

In polymer chains, individual units are held closely together (along the chain) by covalent bonds. So, if each unit is, or contains, a chromophore, conditions may be favourable for resonance energy transfer from one unit to its neighbour, or for excimer or exciplex formation.

Excimers, and especially exciplexes, are very reactive chemically and are often intermediates in facilitating the photosensitised reaction of A with B. The most visible reaction of this type is the effect of the herbicide grammoxone.

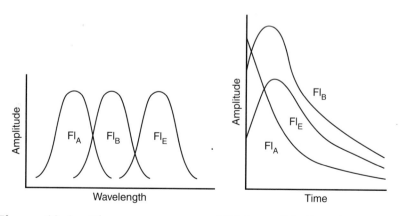

Figure 13.4 *Fluorescence spectra and lifetimes of A, B, and excimer E.*

Grammoxone is one of a family of herbicides based on quaternised bipyridyl compounds.

Under the influence of sunlight, grammoxone forms an exciplex with chlorophyll, which excited state complex then reacts and leads eventually to the destruction of the plant. The advantages of such a herbicide are that it works only on the green chlorophyll containing parts of the plant, and is inactive on stems or in the ground. Further, it works only in sunlight.

13.3   Resonance (Förster) energy transfer

The resonance energy transfer from an initially excited molecule, the donor, to a second molecule, the acceptor, is sometimes called *Förster*, or *Förster–Dexter*, transfer. For it to take place, certain conditions must be fulfilled.

First, as for any resonance phenomenon, the energies of the initial and final states must be equal. So, the process will involve energy levels such that the fluorescence emission energy of the donor equals the absorption energy of the acceptor. If the energies are equal then the light wavelengths will be equal. So, the regions where the energy equality requirement is satisfied are the spectral overlap regions of the emission spectrum of A with either the absorption spectrum of A (for A* to A transfer) or the absorption spectrum of B (for A* to B) transfer (Figure 13.5). The greater the overlap integral, the higher the probability of transfer. Conversely, no overlap means that no transfer is possible (Figure 13.5).

The second requirement is that the donor and acceptor molecules must be in close proximity. The change between the different electronic levels involves a change in the distribution of electrons in the molecule. This geometrical change can be represented by a dipole, called the *transition dipole* for the change.

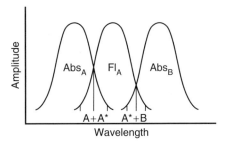

Figure 13.5   *Spectral overlap when energy levels are equal.*

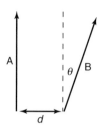

Figure 13.6 *Transition dipoles.*

Then the resonance transfer occurs by way of the transition dipole in the donor molecule inducing a transition dipole in the acceptor molecule. In other words, it is a dipole–induced dipole interaction. Now, readers have met this before when studying Van der Waals forces, one component being able to induce a dipole in a neighbouring molecule. This interaction has a strength that diminishes as the sixth power of the distance between the dipoles. So the probability of resonance transfer also reduces as the sixth power of the separation between the donor and the acceptor. This is a very sensitive dependence on distance.

Thirdly, the two molecules must be favourably aligned. Optimally, the transition vectors of both molecules should be parallel. However, some departure from strict parallelism is allowed and the probability of resonance falls off as the square of the cosine of the angle between the two vectors. So P_{AB} is proportional to $\cos^2\theta/d^6$ (Figure 13.6).

Finally, a word is used to describe this quantum of energy that moves by way of shifting from one molecule to another. Such a mobile quantum is called an *exciton*.

13.4   Exciton movement in polymer chains

If each repeat unit of a polymer chain is a chromophore then the covalent bonds of the chain hold adjacent chromophores close together, so optimising the probability of resonance energy transfer between chromophores. If there is only one type of chromophore in the chain then energy can move from one chromophore to the next in a series of resonance interactions. This constitutes a hopping random walk of the *Förster exciton* along the chain. When this involves the same type of chromophore, the process is called *exciton migration*. However, if the chromophores are different, the movement is called *exciton transfer* (Figure 13.7).

This picture of the energy transfer is simplistic, as in a real polymer the chromophores will have alignments which are controlled by the conformations which the chain may adopt. As discussed in earlier chapters, increasing the temperature may introduce local *gauche* conformations and these will have a different

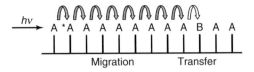

Figure 13.7 *Exciton migration and transfer.*

interaction with their neighbours than sequences which posses a *trans* conformation. The luminescence spectrum exhibited by a polymer reflects the conformational distribution of the excited chain. Thus, whilst B may be another monomer in the same conformation as A, it is also possible that it could be the same monomer but in a different conformation.

This migration along the chain becomes important if the excited chromophore is to react with, or have its energy removed by (be quenched by), a second molecule. The capture radius for reaction of A* with B when A is polymeric is much greater than when A is a small molecule (Figure 13.8).

The migration of an exciton along the polymer chain is much faster than the diffusion together of dilute small molecules in a mobile solvent. Furthermore, the migration is not dependent on translational diffusion (though it is dependent on internal rotation) and so the reaction may well be able to take place in a solid, which is not the case for separated small molecules.

In this example, the polymer chain acts rather like the antenna of a television set, picking up the "signal" of the incident photon and transmitting it to the site of activity. Consequently, the phenomenon has been called the *antenna effect*.

Interestingly, photosynthesis makes use of a similar process. The photons incident on the leaf are "harvested" by stacks of chlorophyll, which then allow the energy to migrate through them to the site of chemical reaction in a slightly different chlorophyll molecule.

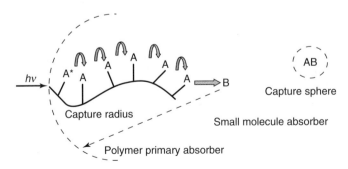

Figure 13.8 *Capture radii for reaction of A* with B when A is polymeric as opposed to a small molecule.*

13.5 Excimer trapping

Unfortunately, the polymer chain geometry that favours exciton migration may also favour excimer formation. Then the existence of an excimer forming site on the chain may create a barrier to the further migration of an exciton. The energy of the incident photon is collected by the absorbing chromophore and the resulting exciton migrates from chromophore to chromophore until it reaches the excimer forming site. Since the energy of the excimer is lower than the energy of the chromophore excited state, the exciton remains on the excimer until it either is lost as excimer fluorescence or is re-energised by thermal energy (Figure 13.9).

Most simple aromatic polymers like polystyrene, polyvinylnaphthalene and poly(N-vinylcarbazole) exhibit both exciton migration and excimer trapping.

13.6 Delocalised (Frenkel) excitons

When the electronic interaction between the chromophores along a polymer chain is so strong that the electrons may be delocalised over several chromophores, the exciton may also be similarly delocalised. Such a delocalised exciton is called a *Frenkel exciton*. Frenkel excitons have been observed in conjugated polymers such as polyphenylacetylene (Figure 13.10).

The almost instantaneous availability of the exciton energy over the whole of the conjugated sequence means that the collection of energy for passage to

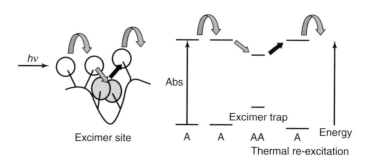

Figure 13.9 *Exciton trapped as excimer and thermal re-excitation.*

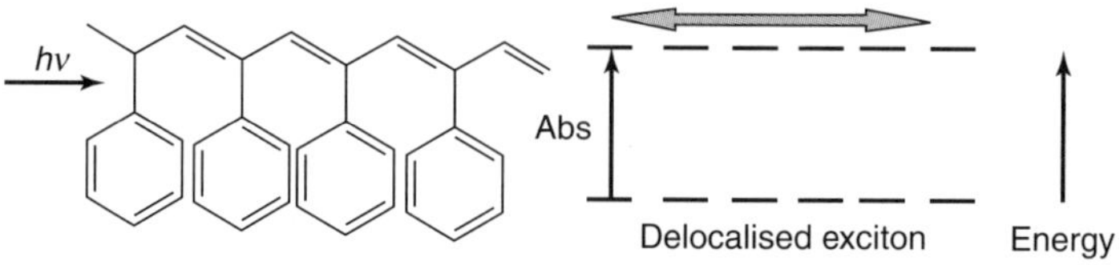

Figure 13.10 *Delocalised (Frenkel) exciton.*

a quenching or reacting molecule is much more efficient than in the case of "hopping" Förster excitons. However, it is important to note that the delocalisation may not extend over the whole length of an alternating double–single bond sequence. This is because the coplanarity required for the conjugation is of very low entropy. Outside a crystal structure, and at some critical length, the requirements of entropy cause some rotation around a single bond, thus breaking the conjugated sequence. Consequently, larger delocalisations are encountered in crystals of a conjugated polymer than in amorphous forms of, or regions in, the same polymer. These excitons will be encountered again in the next chapter, dealing with intrinsically conducting polymers.

The general rule of thumb, then, is that polymeric materials are far more efficient light "harvesters" than small molecules in similar concentrations. Consequently, polymers are far more likely to undergo light initiated chemical reactions, such as oxidation or degradation, than are analogous small molecules.

13.7 Photochromism and molecular motion

A number of molecules have the ability to interact with light, changing from colourless to coloured or dark when irradiated with light. These changes are reversible and form the basis of darkening sunglasses, familiar to most people.

One of the oldest, and the most studied, group of photochromic systems is based on the **spiropyrans** and spirooxazines. An example of a spiropyran system is shown in Figure 13.11.

The spiro form is a colorless leuco dye; the conjugated parts in this molecule are separated by sp^3-hybridised "spiro" carbons. After irradiation with UV light, the bond between the spiro carbon and the oxazine breaks. Then the ring opens, the spiro carbon achieves sp^2 hybridisation and becomes planar, the aromatic group rotates to align its π orbitals with the rest of the molecule, and a conjugated system forms. This has the ability to absorb photons of visible light, and therefore appears coloured. The coloured B form is a thermodynamically unstable (higher energy) form of the molecule. So when the UV is removed, it will return

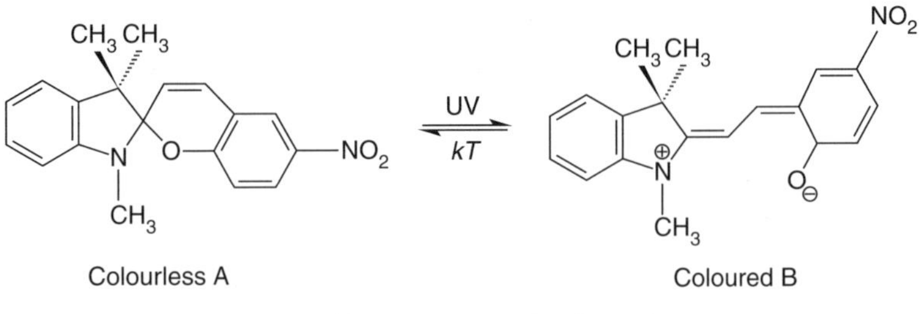

Colourless A   Coloured B

Figure 13.11 *Reversible rotation and colour change in a spiropyran.*

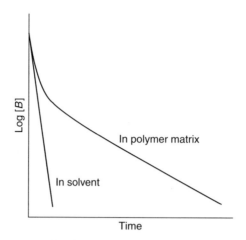

Figure 13.12 *First-order disappearance of the coloured form of a spiropyran.*

to the lower energy A form. This takes place with reformation of the O–C bond and return of the carbon atom to its sp^3 spiro state. Since the conjugation is again limited, the molecule becomes colourless.

These dyes are often dispersed in polymer films. Then the rate at which the coloured form returns to the colourless state depends on the ability of the molecule to rotate inside the polymer matrix. When the molecule is free to undergo internal rotation in a solvent such as methyl methacrylate monomer, the decay is very fast and has a simple first-order dependence (Figure 13.12).

However, when the dye is dispersed in a polymer matrix, the decay of the colour follows more complex kinetics. Initially, the decay appears to be first order and is essentially the same as that observed in the liquid. This is because some of the molecules are located in parts of the matrix where there is sufficient free volume for the rotation process to occur, which is why first-order kinetics, as in a liquid, are observed. After a very short period of time, the decay slows down and becomes non-exponential. This slower decay is associated with molecules which are constrained in the matrix and do not have sufficient free volume to move. So at this stage the colour change depends on the rate at which the molecule can gain enough volume to move. In other words, the decay process depends on the rate at which the matrix can create the necessary free volume. This creation of free volume is essentially the same process as that involved in the glass transition temperature discussed in Chapter 4.

The shape of the kinetic decay curves can be used to determine the distribution of free volume in the polymer matrix. There will be a small number of large holes and a larger number of small holes, giving a typical distribution curve as shown in Figure 13.13.

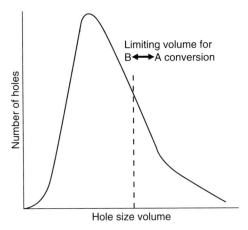

Figure 13.13 *Distribution of hole sizes showing the critical size necessary for photochromic conversion.*

Increasing the temperature will increase the number of holes with a large volume. So the decay becomes faster as the molecules in the excited state, B, convert back to the A form more easily. This phenomenon illustrates the dynamic nature of free volume in a solid polymer; holes will continually grow and disappear as the polymer chains move. Thus the critical parameter for the matrix controlling the rate of decay of the colour is the glass transition temperature.

Further reading

Barashkov N.N. and Gunder O.A. *Fluorescent Polymers*, Ellis Horwood Ltd., Chichester and New York, 1994.

Hoyle C.E. and Torkelson J.M. (Eds.) *Photophysics of Polymers*, ACS Symposium Series 358, American Chemical Society, Washington, 1987.

Kelly J.M., McArdle C.B. and Maunder M.J.de F. *Photochemistry and Polymeric Systems* RSC Publications, Cambridge, 1993.

14

Conductivity in polymer systems

14.1 Introduction

When considering the electrical conductivity of polymers, the phenomenon of interest is the movement of electrical charge and the way that this depends on the chemical structure and motion of the polymer molecules. The charge movement may be electronic or ionic.

Traditionally, polymeric materials have been used as electrical insulators, forming the cladding/sheathing around wires and cables. The insulation characteristics of polymer systems come from the lack of readily ionisable entities and charge carriers. Polymers such as polyethylene and polypropylene form semi-crystalline solids and contain only C–H and C–C– bonds which have relatively high ionisation potentials. As a consequence, the materials exhibit very low conductivities, about $10^{-12}\,\mathrm{S\,m^{-1}}$. Because of the method of manufacture, there will be present a very small number of potentially conducting species, metal catalysts residues, which will usually be trapped in the amorphous regions of the polymer. Nevertheless, the fewer the functional groups present, the more electrically insulating the polymer will be. Materials such as polyesters with polar groups can absorb moisture and exhibit a higher conductivity than alkane based polymer materials. However, in general, all polymers are much more insulating than inorganic semiconductors or metals, for which conductivities range from 10^2 to $10^6\,\mathrm{S\,m^{-1}}$.

14.2 Electronic conduction in polyacetylene

Over the last 50 years, there has been speculation that, with the correct design, it should be possible to make polymers conducting. In principle, if a polymer were to be synthesised such that each of the carbon atoms of the backbone contains a π orbital, then a delocalised electronic structure would be generated. This would mirror early studies of nuclear magnetic resonance spectroscopy showing that the π electrons of the phenyl ring can be polarised and an electrical current induced in them. Extending this idea to a polymer would give a long conjugated polymer chain carrying electrical charge (Figure 14.1).

Indeed, such a polymer chain can be produced by the polymerisation of acetylene to yield a black, very hard solid. The rigidity of the polyacetylene comes

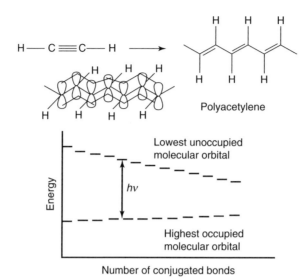

Figure 14.1 *Polymerisation of acetylene to polyacetylene: the energy gap between the highest occupied molecular orbital and the lowest unoccupied molecular orbital decreases with increasing number of conjugated bonds.*

from the conjugated π electronic structure. The energy gap between the bonding and non-bonding (conducting) states for the electron decreases with an increase in the number of bonds involved in the conjugation. Now, the colour which is observed in many organic colouring materials is associated with a conjugated structure having an energy gap appropriate for photo excitation in the visible region. So conjugated polymers, like black polyacetylene, are highly coloured. In such cases, a spectrum of conjugation lengths is available, allowing absorption of energy to occur over the whole of the visible spectrum. The electron is now able to move down the polymer chain and in principle, therefore, could conduct electricity as in a metal wire.

The promotion of a single electron from the highest occupied molecular orbital to the lowest unoccupied molecular orbital produces a *soliton*. The spin state of the excited electron will be conserved during the excitation process and relaxation back to the ground state is a spin allowed transition. For reasons of spin conservation, the free electrons will usually exist as pairs, called *bipolarons*. However, in the promotion of the two electrons to create a bipolaron, the possibility exists that the spin of one of the electrons changes (Figure 14.2).

The lifetime of the excited state will depend on the spin state and a bipolaron with the spins unpaired will have a longer lifetime than that with spins paired. Studies of polyacetylene indicate that as the temperature is changed, so the dominant species changes. At higher temperatures conduction involves predominantly bipolaron states.

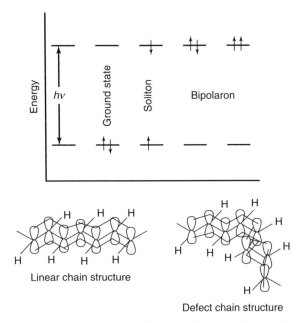

Figure 14.2 *Schematic representation of the ground state and excited electron spin states for an intrinsically conducting polymer: conformation of a chain corresponding to a defect structure.*

The energy to promote the electrons to the lowest unoccupied molecular orbital can be provided by either photons, which is photoconductivity, or by phonons in the lattice, which is thermoconductivity. In the case of polyacetylene, the energy gap is sufficiently small for phonons to be able to promote electrons to the conduction state.

Clearly, injection or removal of an electron from the conjugated backbone is essentially a chemical oxidation or reduction process. So polyacetylene is sensitive to oxidation by atmospheric oxygen. Oxygen can add to the polymer backbone and destroy the π conjugation, which in turn reduces the range of delocalisation of the electrons. This leads to a reduction in the conductivity of the polymer.

Whilst the electrons can move freely along the polymer chains, it is still necessary to put electrons into, and take electrons out of, the polymer. Making polymers intrinsically conducting is done by doping the material. A variety of dopants have been used. These have the ability to either inject electrons into, or abstract electrons from, the conjugated backbone. Adding or removing an electron from the conjugated backbone is a chemical oxidation or reduction. So it is not surprising to find that polyacetylene is sensitive to oxidation by atmospheric oxygen. Oxygen can add to the polymer backbone, thereby destroying the π conjugation and reducing the conductivity.

In terms of polymer dynamics the chains are very rigid, but defects can arise because the backbone does not adopt a low entropy all-*trans* structure. Part of the temperature dependence of the conductance of these materials is associated with the creation of such defects. Increasing temperature will increase the number of defects, converting the *trans* structure into a *gauche* form for which the energy of the lowest unfilled molecular orbital is slightly higher than that of the all-*trans* state. This reduces the distance over which the free electron can move and hence the conductivity. In addition, non-linearity of the chain influences the ability of the chain to interact with lattice phonons, also reducing the probability of an electron achieving a conducting state.

14.3   Electronic conduction in carbon nano tubes

Polyacetylene can be considered as a linear, one-dimensional, conductor. To achieve longer range electron mobility, it is desirable to construct a two-dimensional conductor in which the π electron cloud extends in a plane. To visualise this, consider a number of polyacetylene chains laid alongside each other, and then replace the carbon to hydrogen bonds on each chain by interchain carbon to carbon bonds. Such sheets are made up of only carbon atom and are relatively inert to oxidation. If the sheet is present as a simple planar structure then it is called *graphene*, which is an exfoliated form of graphite (Figure 14.3).

These sheets can be rolled up to form tubes. Bending a single sheet then forms a single-walled nano tube (SWNT). If more than one sheet is incorporated then we have multi-walled nano tubes (MWNT). In the simplest form, the hexagonal benzenoid rings connect up to form a tube in which the rings are formed with the same base plane and are aligned in the tube direction, an $(n,0)$ zigzag structure.

However, it is also possible that the sheet can be rolled so that the transverse axis lies along a plane diagonal which can be defined by $C_h = na_1 + ma_2$, where a_1 and a_2 are the unit vectors of the base graphene sheets. At first sight one might expect the choice of the alignment of the primary rings would have little effect on the tubes which are formed. However, the way in which these sheets are rolled can give different electrical properties. If $n = m$, the rings match perfectly and the nano tube is metallic. However if $n - m$ is a multiple of 3, then the nano tube is semiconducting with a very small band gap. The observed differences in conductivity reflect the strain on the rings. The distortion of the bonds out of planarity will tend to localise the electrons and increase the effective energy gap associated with promotion of the electron from an occupied to an unoccupied level.

In theory, metallic nano tubes have a greater current carrying capacity than copper but, as with polyacetylene, the conductivity ultimately depends on the ability to inject and remove electrons So it is limited by the dopants or the chemistry of the electrode to which the tubes are interfaced.

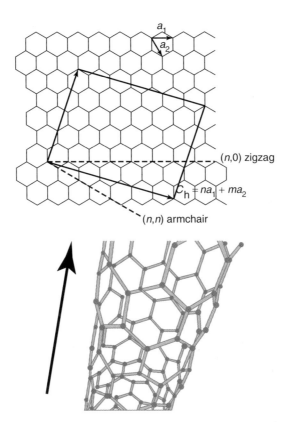

Figure 14.3 *Schematic representation of the basal structure of graphene; a_1 and a_2 are the unit vectors and structure for an (n,0) zigzag nano tube; the arrow indicates the alignment of the directional vector.*

14.4 Polypyrrole, polythiophene and related systems

In an attempt to achieve polymer systems more stable than polyacetylene, a range of polymers having a conjugated polymer backbone have been synthesised. Substituting the hydrogen atoms in polyacetylene gives a range of materials in which the conjugation can be incorporated into a five-membered or six-membered ring (Figure 14.4).

High intrinsic conductivities have been observed for all these polymers. Because they are more stable to oxygen than polyacetylene, they have proven to be more useful in various technologies.

Polypyrrole has the interesting property of having a conductivity which depends on the pH of the medium. In an acid, the N–H group can be quaternised into the $-N^+H_2$ form. The protonation can change the conductivity of the polymer by four orders of magnitude, so allowing it to be used as a pH sensor.

The related polymer, polyaniline, can exist in several forms depending on its state of oxidation or reduction. The oxidation and reduction of the polymer

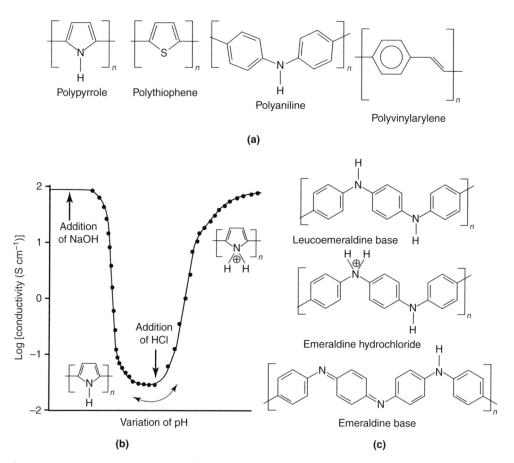

Figure 14.4 *(a) Structures of some intrinsically conducting polymers: polypyrrole, polythiophene, polyaniline and polyvinylarylene; (b) change of conductivity with pH for polypyrrole; and (c) various forms of polyaniline.*

changes the colour of the material and this is reflected in the names used to designate the various forms. As in the case of polypyrrole it is possible to obtain a quaternised form, as emeraldine hydrochloride.

Polythiophene has a narrow band gap and electrons can readily be excited by photons, so allowing the polymer to exhibit photoconductivity. Because of this, it is finding applications in solar cells and organic transistors.

Poly(p-phenylenevinylene) (PVA) and its soluble derivatives have photochemical and electrochemical properties which make them useful for electroluminescent as well as semiconducting applications. The modified PVA (PVA plus cyano-PVA), which is currently used in organic light emitting diodes, is shown in Figure 14.5.

Electrons are injected into the polymer via an indium tin oxide transparent conducting electrode. The other electrode is usually formed from a thin

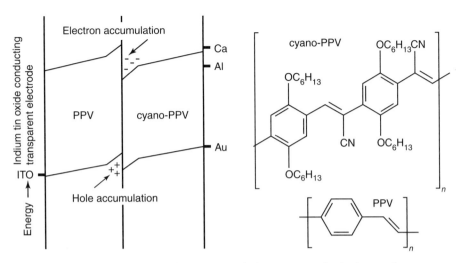

Figure 14.5 *Layered structure for a organic light emitting diode device showing variation of potential across the device and the bound hole trap which produces the light emission.*

film of deposited calcium. The charge accumulation is shown at the junction between the PVA and the cyano-PVA where an energy gap (difference) exists. Relaxation of the charged state leads to emission of a photon and is the basis for the very thin displays currently available in flat screen televisions and monitors. Whilst these materials, the conducting polymers and the calcium electrodes, are still sensitive to oxygen and have to be carefully protected from the atmosphere, they are being successfully incorporated into a range of electronic devices.

In general, active polymer chain dynamics do not have a very major effect on the properties of these materials. However, in the case of PVA the nature of the excited state from which luminescence occurs depends on the chain alignment and folding that has occurred during preparation. This static conformation variability produces defects similar to those found in polyacetylene. This limits the range of conjugation and so significantly influences the efficiency of light emission.

14.5 Temperature sensitive electrical conductivity

The dynamic movement of polymer chains has been used to achieve materials with a *positive temperature coefficient* of resistivity: PTC materials. The characteristic of these materials is that the resistance does not decrease with increase in temperature, as is observed with most semiconductors, but increases at some characteristic temperature T_p (Figure 14.6).

These materials are fabricated using conducting carbon black, graphite whiskers or metallic fibres as filler.

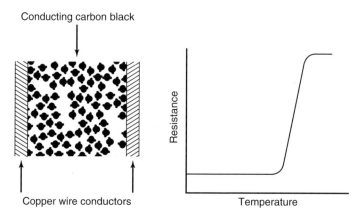

Figure 14.6 *Schematic representation of a positive temperature coefficient material and a typical plot of the variation of resistance with temperature.*

For example, if carbon black is added to an insulating polymer matrix such as polyethylene, a point will be reached at which the particles can form chains of touching particles. At this point, electrons can jump between individual particles and conductivity is achieved. However, if such a structure is heated, the polymer chains will expand and the matrix will open. The net effect is that contacts between the particles will decrease and the electrical conductivity will be diminished.

The largest expansion of the solid occurs at the melt flow transition of the polymer. The molten polymer can flow so that the contacts between the carbon black particles are not recovered on cooling. To avoid melt flow, it is usual for the material to be lightly cross-linked, as was described in Chapter 7 for the behaviour of rubbers. This prevents irreversible flow, but allows the large expansion of the matrix required to change the electrical conductivity. This concept is used in the fabrication of heating tapes, which can be wound around pipes and apparatus to produce distributed heating, but also allow control of the temperature.

14.6 Polymeric electrolytes

Batteries require the transfer of ions from one electrode to another. Originally in battery technology, the electrolyte was a fluid. However, batteries for uses in space craft and medical applications, such as cardiac pacemakers, require that the electrolyte should be a permeable solid. The original pacemaker batteries used a mixture of poly(*N*-vinylpyrrolidone) mixed with a low molar mass plasticiser and doped with a high concentration of a salt.

Modern lithium ion batteries use poly(ethylene oxide) or polyacrylonitrile as the solid state electrolyte. Ideally, the lithium-salt electrolyte is dispersed in a solid polymer composite, but ion diffusion will occur only if the polymer is heated above its melt temperature. In poly(ethylene oxide) electrolytes, there is still some controversy as to whether the lithium ion migration occurs in the disordered amorphous phase or whether some ions can move within the crystalline region. Poly(ethylene oxide) adopts a helical crystal form. The oxygen atoms face into the axis of the helix and so it is possible that the metal ion can be coordinated by the oxygen, rather as inside a crown ether molecule. Whilst this mechanism cannot be discounted, the increased conductivity at the melting point of the polymer indicates that the increased disorder, with greater available volume, allows greater mobility of the lithium ions.

Current batteries are hybrid structures in which low molar mass solvents are added to increase the amorphous content and decrease the melting temperature. The batteries are constructed by rolling up sheets of electrodes with the polymer electrolyte as a separator (Figure 14.7).

Because of the large area of electrodes so obtained, it is possible to achieve the current density necessary to produce a useful battery. The cathode is usually constructed by impregnation of carbon films with $LiCoO_2$ or $LiMnO_4$ and the anode is a carbon–lithium intercalation compound.

The charging process requires lithium ions to be transferred between the electrodes and this takes place through the polymer electrolyte, which also prevents short circuiting of the electrodes. The battery reaction can be summarised as:

- *anode*: $carbon–Li_x \rightarrow C + xLi^+ + xe^-$
- *separator*: Li^+ conduction
- *cathode*: $Li_{1-x}CoO_2 + xLi^+ + xe^- \rightarrow LiCoO_2$

The polymer electrolyte separators are usually poly(ethylene oxide) plus a suitable lithium salt as described above. Intrinsically conducting polymers, such

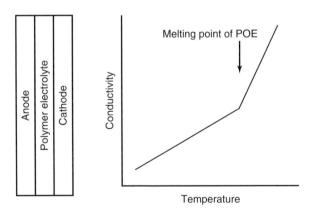

Figure 14.7 *Schematic representation of a solid state battery and the temperature variation of conductivity.*

as polyaniline, can be used as the electrode materials in certain applications. Currently the development of better batteries is a continuing activity, very much driven by the automotive industry. Progress relies on an understanding of the intrinsic conductivity of polymeric electrode materials and of the mobility of ions in "solid" polymer electrolytes.

Further reading

Kuzmany H., Mehring M. and Roth S. (Eds.) *Electronic Properties of Conjugated Polymers*, Solid State Science 76, Springer Verlag, Berlin, 1987.

Salanecki W.R., Lundstrom I. and Ranby B. (Eds.) *Conjugated Polymers and Related Materials*, Oxford Science Publications, Oxford, 1993.

15

Diffusion in polymers

15.1 Introduction

While the main subject of this book is the movement of polymer molecules themselves, in the chapters dealing with electrical and photo phenomena, we have already met the movement of other entities inside polymers. Very often the movement of these species is coordinated with, or controlled by, chain movements in the polymer matrix. The same is true of the diffusive motion of small molecules into, inside and through polymers. Such phenomena are important in a wide variety of uses ranging from packaging, through membrane separation processes, to controlled drug release.

It is important to distinguish between *permeability, P* and *diffusivity, D*. The former is a measure of the amount of material that can be absorbed into one side of a polymer sample and then extracted from the other side. This penetration of a small molecule species into a bulk polymer is called *permeation*. On the other hand, the statistical random walk of sorbate molecules inside the polymer is called *diffusion*. Another important term is *solubility, S*. Whilst the ability of a molecule to move through the matrix is clearly important, its solubility can also be a significant factor in defining the permeability. In fact, the permeability is directly related to the product of the diffusivity and the solubility:

$$P = SD$$

The movement of smaller molecules into and through a polymer depends significantly on two phenomena. The first is the nature of the interactions between the diffusing molecule and the polymer chain, as evidenced macroscopically in the solubility or miscibility of the two. The second is the availability of free volume into which the diffusing species can move. For this reason, diffusivities differ markedly between different polymer types and also between the crystal, glass and rubber forms of any one polymer. A useful rule of thumb is that the lower the available free volume, the slower the diffusion, but the greater the selectivity of the transport process.

Many of the technological uses of these molecular processes depend on the permeability of the sorbate–polymer system. Permeability measures both the amount of solute molecules in the polymer and the speed with which they can move, as was presented above. Looking at the effect of temperature, as was done for polymer chain motions in Chapter 4, this is incorporated into a composite "permeation activation energy", which is the sum of the enthalpy of solution and the activation energy for diffusion. Because molecules of like chemical structure are usually more miscible than dissimilar ones, another good rule of thumb is that the permeation of molecules in a polymer is highest when the two are chemically similar.

The effect of similarity and dissimilarity is evident in the comparison of non-condensable gases with liquid or even solvent molecules. The former have little effect on the polymer matrix and so the diffusivity is little affected by the amount of gas absorbed. On the other hand, condensable or solvent species can open up the matrix, when the diffusivity becomes very much a function of the amount of absorbed material present. In such cases, the permeating molecules function rather like the plasticisers discussed in Chapter 4.

15.2 Mechanism of diffusion

The frequency of movement of a sorbate molecule trapped between polymer chains is much higher than that of the chain segments themselves. Consequently, it is the polymer motion, permitting a jump to a new site, that is rate determining. This movement requires both activation energy and a critical activation volume, as we saw in Chapter 4. Further, the sorbate molecule can move more rapidly parallel to surrounding chain backbones than perpendicular to the backbone directions. This makes the perpendicular transport the rate determining step. Then the free volume necessary for this can be formed only by the segmental motion of the entrapping chain units.

The role of solubility in permeation is of prime importance. One aspect of this is especially significant for diffusion in a glassy polymer. The non-equilibrium nature of the glass means that there is a distribution in the size of the free volume spaces between the polymer chains. Some of these are so large that they can be called "voids". These can hold several trapped sorbate molecules. Obviously, the presence of these voids depends on the nature of the polymer and its processing prehistory. Then diffusion occurs by movement of single molecules "dissolved" in the polymer, but the concentration gradients that drive the direction of diffusion are perturbed by the presence of these reservoirs of trapped sorbate. As a result, observed absorption and diffusion isotherms for permeation in a glass are complex and difficult to interpret.

15.3 Permeation of permanent gases

In amorphous or rubbery polymers, a permanent gas forms only a very dilute solution in the polymer. Consequently, there is no distortion of the polymer matrix by the gas and the diffusion coefficient is independent of the amount of absorbed gas. Then the permeation follows the ideal product of solubility with diffusivity rather closely. If the temperature dependence of permeability is given by the exponential Arrhenius type equation

$$P = P_0 \exp(-\Delta E_{\mathrm{p}}^{\ddagger}/RT)$$

then

$$\Delta E_{\mathrm{p}}^{\ddagger} = \Delta H_{\mathrm{s}} + \Delta E_{\mathrm{D}}^{\ddagger}$$

where $\Delta E_{\mathrm{p}}^{\ddagger}$ is the activation energy for permeation, ΔH_{s} is the enthalpy of solution and $\Delta E_{\mathrm{D}}^{\ddagger}$ is the activation energy for diffusion.

In a homogeneous substrate the diffusivity and permeability are isotropic. However, the polymer may have been orientated during processing. If that should be the case, permeation is greater along the orientation dimension than perpendicular to it.

It has been found that polymers with bulky groups on the backbone or in the side chain have lower permeabilities than sterically less hindered materials.

The situation is much more complex for semi-crystalline materials. Generally, diffusion in a perfect crystallite will be negligible. Then the diffusion path wanders around and between crystallites. It is thus lengthened and the overall permeability decreases. So, when the amorphous content is above 30% to 40%, the permeability and diffusivity increase with increasing amorphous content. However, the crystallites are not perfect, but contain defects within which sorbate gas movement is enhanced. The number of defects increases as the crystallinity increases above 70% to 80%, so the permeability increases as the amorphous content is decreased below about 35%. In other words, the permeability is a minimum at about 65% crystallinity (Figure 15.1).

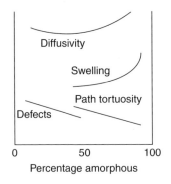

Figure 15.1 *Defect concentration, path length tortuosity, swelling and resulting diffusivity for vapour in a semi-crystalline polymer.*

In both the homogeneous amorphous phase and the crystal defects, the gas diffusivity is a function of molecular size. Obviously, the smaller molecules have a greater diffusivity than the larger ones. However, the critical size factor, or selectivity, is different in the two phases.

15.4 Permeation of condensable vapours

Permeation of vapours which are condensable or which interact strongly with groups on the polymer chain does not follow the simple treatment outlined above. There are two important reasons for this. Firstly, the presence of sorbate affects the chain segment jump process. Thus the diffusion coefficient is not independent of the concentration of diffusing species. Secondly, as the temperature is raised, the diffusion coefficient increases, but the solubility, if it follows Henry's law, decreases. Then at temperatures above the glass transition temperature of the polymer these two factors balance each other and the permeation does not increase as much as might be expected.

Generally, there are two types of permeation. Above the glass transition of the polymer the rate of segmental relaxation is greater than the rate of sorbate penetration. Then the permeability is little affected by sorbate-induced swelling. So the diffusivity is not very concentration dependent. This is the situation with many hydrocarbon sorbates in commercial rubbers. On the other hand, below the glass transition temperature the segmental relaxation rate is much less than the diffusivity, and so in these cases the permeability is very dependent on swelling factors.

Of particular technological importance is the permeation of water. Here the ability of the water molecule to form hydrogen bonds either with groups in the polymer or with other water molecules has an enormous effect. In the former case, in hydrophilic polymers, the water exerts a swelling, plasticising effect. The diffusivity increases with penetrant concentration. However, in hydrophobic polymers, the water tends to bond to itself, forming clusters of water molecules. This decreases the amount of water diffusing through the polymer, and so the diffusivity decreases with overall concentration. Using these two effects, it is possible to synthesise polymer films which have controlled barrier permeation properties.

15.5 Some technological applications

Having discussed the permeation of water above, it is appropriate to start with an application depending on control of water diffusion. This use is in dressings for covering burn and scald wounds. The material covering the burn must permit a controlled transit of water, but must provide a barrier against bacterial and other infections. This is achieved by the use of partially swollen hydrogels formed from cross-linked hydrophilic polymers. The small water molecules have a high diffusivity, while the large organisms cannot penetrate the film. Indeed, antibiotic substances can be incorporated into the film, further aiding its protective properties.

Another medical use is the controlled release of drugs. When a drug is administered orally or by injection, the concentration in the body rises rapidly to a high value, then drops in what is usually an exponential manner. So, in order to have an effective concentration over a reasonable time, the initial maximum may be undesirably high. Alternatively, to keep the initial concentration at a safe level, the longer term concentration may become too low to give the desired effect. This problem can be overcome by incorporating the drug into a polymer through or from which it can diffuse at a planned rate.

One mechanism for doing this is to use a *transdermal patch*. This is essentially an adhesive plaster with the drug mixed into the adhesive. Polyisobutylene adhesives are often used for this purpose. The drug then diffuses out of the polymer and through the skin at a preplanned rate. Of course this only works for drugs that can be absorbed through the skin.

An alternative method is to enclose the drug in a small capsule that can be implanted in some way. The drug then diffuses through the polymer skin of the capsule and into the body. For low doses over a long time the capsule may be placed subcutaneously by a small operation. Alternatively, or for more localised treatment, the capsule may be inserted as a suppository. Silicones such as polydimethylsiloxane are widely used for this purpose.

Although these medical uses are exciting, the largest tonnages are for membrane separation and packaging processes. Semi-permeable membranes have been used over many years for the purification of sea water by reverse osmosis. A hydrophilic polymer membrane allows the passage of small water molecules, but not the sodium and chloride ions surrounded by their hydration shells. Then application of pressure to the brine side overcomes the osmotic pressure of the solution, forcing water through the membrane into the low pressure purified side. The pore size for this permeation is controlled by the extent of swelling and cross-linking in the polymer. Modified cellulose acetate membranes swollen with 20% to 30% by volume of water are widely used. At the same, time the membrane must support the pressure difference, which the swollen separation film cannot. This is achieved by supporting the permeation film on an open, strong network of some hydrophobic polymer.

Rather similar concepts are applied in the use of polymer membranes for separation and other purification processes. The list of polymers used in separation membranes is very long. As well as widely available homopolymers, there are many copolymers specially "tailored" to give the desired separation properties. However, two general aspects are important. The first is the nature of the intermolecular interactions between the polymer and potential sorbates. Thus hydrophilic polymers are used to permit the passage of water and polar substances while restricting the permeation of non-polar materials. Conversely, hydrophobic polymers have the opposite effect. The second is the balance between cross-linking and swelling in determining the "pore size" of the membrane so that selectivity can be based on the molecular volume of the sorbate.

One use that is of increasing importance is in the treatment of waste water from industrial processes. Not only is this important for the environmental quality of the discharged water, but it also allows recovery of substances that may be of value to the processor. As in reverse osmosis, the film must transmit water molecules, but stop organic or ionic impurities.

While separation processes in liquid media usually involve swollen polymer membranes, this is not the case with separation of gases. Here the smaller molecular cross-section of the diffusing species requires a more constricted diffusion path. Nowadays membrane separations are widely used in industrial gas plants to separate oxygen and nitrogen in air, and in refineries to separate carbon dioxide and methane. For these purposes many different kinds of polymer are used, with heavy emphasis on polyolefins. Often the discrimination is obtained by using polymers below the glass transition temperature. Here, though, there is sometimes a problem caused by the non-equilibrium nature of the glassy state. Then the membrane properties can change with time and with handling.

When polymer membranes are used in packaging, the objective is to prevent, not to facilitate, permeation. Gases, vapours, liquids and organisms all must be stopped from reaching the wrapped article. When a packaging film is used to protect foodstuffs, not only must oxygen be prevented from diffusing in, carbon dioxide must be prevented from diffusing out. This two way diffusion of different substances is called *counter diffusion*. So, polymers are chosen which have the most restricted diffusivity while possessing the required mechanical properties.

An interesting example is the use of plastic bottles to contain pressurised carbonated drinks. This is usually done with microcrystalline poly(ethylene terephthalate), PET. This polymer gives good mechanical strength, retaining a measure of flexibility. PET has microcrystallites which are approximately 10 nm in dimensions. These crystallites are linked by more amorphous chains. Permeation is restricted to the amorphous regions. For example, bottles for carbonated soft drinks are produced by a blowing process which stretches and aligns the crystallites parallel to the walls and increases the tortuous path for the diffusion of the carbon dioxide molecules through the polymer. The result is a large bottle in which the crystallites are highly drawn and well aligned, so have better barrier properties than smaller bottles which have less well aligned polymers. The carbon dioxide pressure falls slowly as the gas is first absorbed into, then permeates through, the bottle. However, the rate of this is so slow that the pressure drop is not more than about 10% per month, giving excellent shelf life for the product.

15.6 Diffusion controlled homogeneous polymer reactions

Most chemical reactions require two reactant molecules to come into close proximity before they can react with each other. If the reactants are not initially in contact then the overall process involves two steps. First, the reacting molecules must move to be next to each other, and then chemical change may, or may not, occur.

Many reactions in heterogeneous systems, such as at interfaces or in flow, are controlled by the rate at which the reactants can come together. However, there are also situations in homogeneous systems where diffusion is important.

For the majority of reactions carried out in the gas phase or in a liquid, the molecules come together many times before the chemical change takes place. The rate of the reaction is then controlled by the rate of the chemical step. However, if a very fast chemical reaction is involved, or if the reaction takes place in a solid or semi-solid medium, the approach diffusion step may be much slower than the chemical reaction. Then the overall process can occur only as fast as the molecules can come together. Any such reaction is said to be *diffusion controlled*.

Many reactions of free radicals, of electronic excited states and of oppositely charged ions are found to be diffusion controlled in condensed phases. When these reactions involve polymeric reactants, or take place in a polymer medium, the diffusion step will be dependent on the nature of the polymer molecular motions in the reaction zone.

15.7 Diffusion control in polymerisation

The role of diffusion in controlling the rate of polymer reactions is best illustrated by free radical polymerisation of a liquid monomer. The overall rate of polymerisation of liquid monomer to solid polymer is exemplified by the free radical polymerisation of methyl methacrylate at 30 °C (Figure 15.2).

Starting with liquid monomer, the polymerisation rate slowly decreases as monomer is consumed to form polymer. Then a very significant auto-acceleration occurs, raising the rate many fold. Since the polymerisation is exothermic,

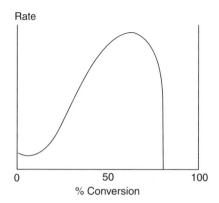

Figure 15.2   *Rate of bulk free radical polymerisation of methyl methacrylate at 30 °C.*

unless precautions are taken to keep the reaction at a steady temperature, the extra heat generated can increase the temperature, thereby accelerating the rate even further. Indeed, a runaway reaction leading to thermal explosion can occur. This is called the gel effect, or often the Norrish–Tromsdorff effect. As the reaction passes the halfway mark, the rate starts to fall and eventually drops to zero. In the case of methyl methacrylate, this occurs before all the monomer has been converted to polymer (100% conversion). So for this, and many other monomers, the polymer obtained contains residual monomer, a highly undesirable characteristic.

Free radical polymerisation is a chain reaction with three main steps. These are formation of free radicals, called *initiation*; the growth of a polymer chain by successive addition of monomer molecules to a free radical at the end of the growing chain, called *propagation*; and cessation of radical growth, usually by mutual destruction of two free radicals, called *termination*. The rate of polymerisation is a balance between these three steps, with initiation and propagation increasing the rate and termination decreasing it. Then the auto-acceleration and final drop to zero stem from changes in the ratio of these steps as a result of diffusion control of the termination and propagation reactions respectively.

15.8   Diffusion controlled free radical termination

Early explanations of auto-acceleration postulated that at the onset point the viscosity of the polymer in monomer solution became sufficiently great that growing polymer radicals could not diffuse together in order to terminate each other. The decreased termination rate then led to the increased overall polymerisation rate. However, it is now known that the radical–radical reaction is diffusion controlled from the very start of the reaction, although the large acceleration in rate does not occur until a considerable amount of polymer has been formed.

When the reacting centre, here a free radical, occupies only a small volume element on the chain, two different diffusive molecular motions are involved in bringing two reactants together. First, two chain radicals initially separated by non-radical species must diffuse together. This is translation and of course occurs by way of the reptation process described previously. However, two chains can be in proximity to each other while the reactive chain ends are still separated by other chain segments or monomer molecules. So, a conformation rearrangement of the chains is necessary to bring the two radical chain ends into contact. This conformation change is the segmental rotation, also described earlier. The overall reaction of two macro-radicals can then be represented by the following three step scheme:

$$A° + B° \xrightarrow{\text{Translation}} °AB°$$

$$°AB° \xrightarrow{\text{Segmental rearrangement}} A°°B$$

$$A°°B \xrightarrow{\text{Chemical step}} \text{Inert polymer}$$

where $°AB°$ represents two chains in proximity but with the radical chain ends widely separated, and $A°°B$ represents the same chains with the ends rotated into contact.

It transpires that in those systems that have been studied it is the segmental rearrangement process that is rate determining in the early stages of the reaction. So polymer chains with large steric barriers to backbone rotation have lower termination rates than polymer chains with unrestricted rotation. Turning again to the polymerisation of methyl methacrylate, at the start of the reaction there are relatively few inter-chain entanglements and the chain end rearrangements are unhindered by such. However, as more polymer is formed, there are more and more entanglements, the viscosity rises, the chain rotations become more hindered and so the termination rate decreases. In the final stages of polymerisation, the growing chains become so entangled in the polymer network that translation virtually ceases. Then the chain end becomes a *trapped radical*. It can change its location only by adding monomer and so growing into a new region of space.

15.9   Diffusion controlled propagation

In the propagation step a monomer molecule adds to the free radical end of a growing chain and in so doing generates another radical. This process is fast by comparison with many chemical reactions, but it is not fast enough to be diffusion controlled under normal circumstances. However, as the polymerisation proceeds, the monomer molecules become used up and so become further apart. Also, the viscosity of the monomer–polymer mixture increases until at the end of the reaction it is solid polymer. In these final stages, the translational diffusion of monomer molecules to the radical centres of reaction becomes restricted so that the reaction becomes diffusion controlled, slows, and finally stops.

An interesting question is: why does the reaction stop before 100% conversion of monomer to polymer? The answer to this is found by carrying out the later stages of the reaction at higher temperatures. At progressively higher temperatures, the achievable final conversion increases. It finally becomes 100% at a temperature that corresponds to the glass transition temperature of the solid polymer!

If we work back from solid polymer by adding progressively more amounts of liquid monomer, we see that the glass transition temperature of the now plasticised polymer decreases with the amount of monomer added. Monomer cannot diffuse through polymer below the glass transition temperature, so the reaction stops when the amount of monomer decreases by the amount necessary to raise the transition temperature of the mixture above the reaction temperature. However, the reaction will continue to proceed if the temperature of the reaction is raised, and will go to 100% conversion if the final temperature is above the transition temperature of the solid polymer. For this reason, many industrial processes carry out the polymerisation using a temperature profile that finishes with a high temperature to ensure that there is no unreacted monomer left in the final product.

15.10 Polymer chain end excited state quenching

Another interesting case where the reaction of polymer chain ends is diffusion controlled is found when a chain contains an active chromophore at one end and a quenching agent at the other. Then quenching of the excited state of the chromophore occurs when the two reactants on the chain ends come together and react as described in Chapter 12. This quenching reaction is extremely fast, occurring almost every time that the reactants come into proximity. So the rate determining process is diffusion. In this case, the two reacting species can be on the same chain or on different chains.

The intrachain quenching shows that, just as in the termination of radical chain ends, the rate of this process depends on the rate of segmental rearrangement of the chain. So again the quenching reaction rate is a measure of the chain conformation change rate.

Further reading

Diffusion

Neogi P. *Diffusion in Polymers*, Marcel Decker, New York, 1996.

Vieth W.R. *Diffusion in and Through Polymers: Principles and Applications*, Hanser Gardner Publications, Cincinnati, 1991.

Diffusion controlled reactions

North A.M. *Diffusion Control of Homogeneous Free Radical Reactions*, in J.C. Robb and F.W. Peaker (Eds.), *Progress in High Polymers Volume 2*, Heywood Books, London, 1968.

16

Methods of studying molecular motion

16.1 Introduction

Methods of studying molecular motion in polymers have three principal objectives. The first is to ascertain what are the states of the molecules, interchange between which constitutes the motion. The second is to observe the rates of this interchange movement. The third is to evaluate the relative energies of the states and of the barrier between them. So studies fall into two categories: measurement of the distribution of populations between various conformational states, and the dynamics of interchange between these states.

Some techniques for examining materials give evidence that motion has occurred, but do not observe that motion directly. A good example of this is X-ray diffraction. We saw in Chapter 9 that drawing of semi-crystalline polymers induces a phase change in the crystalline lamellae at the second yield point. This is proved by the X-ray scattering patterns. Obviously, if the molecules have taken up new positions in the crystal, some sort of motion must have occurred. However, the X-rays do not show how this is happening. So such *post hoc* methods are not considered further in this chapter.

Before considering the methods available for studying molecular motion, it is appropriate to remind ourselves of the timescales which the various techniques can access. In Chapter 4, we showed that the frequency of molecular movement and the observation frequency are equal at the transition

$$\omega_t = 1/\tau$$

where ω_t is the observed transition frequency in radians per second and τ is the molecular relaxation time in seconds. So, particular processes can be observed when the inverse relaxation times correspond to the frequencies of observation. Figure 16.1 illustrates the technique wavelength, frequency and timescale of observation.

As so often in polymer science, techniques developed for the study of small molecules are applied to polymer solutions and solids. The methods available use photons, electrons, or neutrons as probes and usually require times of the order of 10^{-3} s or greater to collect the data. On this timescale, the molecules may very well have undergone considerable movement. So it is unlikely that normal

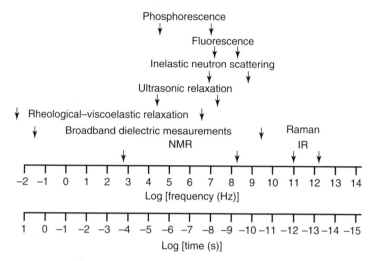

Figure 16.1 *Summary of frequency ranges and timescales accessible by various techniques.*

spectroscopic and scattering methods can observe the time resolved motion of a single molecular unit. Nevertheless, there is information to be gleaned from time averaged observations. Furthermore, there are specialised techniques able to observe group orientations even on a timescale of the order of 10^{-12} s. In this chapter, a summary look at some of the methods used will be presented.

16.2 Spectroscopic methods

Infrared and other spectroscopic methods usually measure the distribution between a small number of possible conformations. Different conformations can have different atomic vibrational energies, so the frequency of a C–X stretch associated with the *trans* conformation may differ from that of the *gauche* form. So the relative intensity of the two infrared bands reflects the equilibrium distribution of the molecules between the two conformations. Changing the temperature will alter this ratio, reflecting changes in the population of each state. Then application of the Van't Hoff isochore allows evaluation of the energy difference between the conformations.

Examining in more detail the conformation change occurring by rotation about backbone bonds, we find that the process can actually give rise to a quantised transition – a *torsional vibration*. This torsional frequency will usually lie in the far infrared range with wave numbers 60–200 cm^{-1}.

This resonance absorption will have a significant intensity only for the lowest energy conformer. In other words, it is looking at an oscillation at the bottom of the potential energy well. If it is assumed that the barrier to internal rotation has a sinusoidal form then it is possible to estimate its height from the frequency of the torsional oscillation. Some of the data obtained from such resonance spectroscopic measurements do compare well with those from relaxation studies. However, since only the lowest energy states are involved, such calculations often underestimate the actual barrier height involved in processes like the glass to rubber transition.

Raman spectroscopy, although it has different selection rules, similarly portrays atomic vibration frequencies that can be sensitive to conformational changes. The infrared and Raman transitions take place in times that are of the order of 10^{-12} to 10^{-13} s and so essentially freeze the distribution on this timescale, i.e. they show the nature of the interchanging states and the energy difference between them, but not the actual interchange rates.

Very similar to Raman spectroscopy is *Brillouin scattering*. Again, monochromatic visible light is used to irradiate a sample. However, instead of the photon energy being transferred to atomic vibrations, it is transferred to larger molecular movements. In the case of solids, the movements between a number of molecules become correlated and are quantised as *phonons*. The phonon spectrum gives information on the kinetics and energetics of these cooperative molecular movements. So, the technique can be thought of as observation of hypersonic relaxation.

The shift in energy from that of the incident photons is very small, being of the order of gigahertz in terahertz. So this technique only became possible with the development of very monochromatic laser sources and high resolution interferometric analytical instrumentation. The relaxation times, fractions of a nanosecond, observed in Brillouin scattering measurements are shorter than those seen in normal ultrasonic methods. So, as was seen in Chapter 4, the high frequency, high transition temperature observations lie in the region where the temperature dependence follows the Arrhenius relationship, allowing simple evaluation of the interchange activation energy.

Another spectroscopic technique that is sensitive to conformation change is *nuclear magnetic resonance spectroscopy* (NMR). This is one of the most powerful techniques for the characterisation of chemical compounds. Its importance for chemists lies in the fact that the characteristic resonance frequency of a nucleus depends on its chemical environment. Different atomic neighbours shield the nuclei under study from the external magnetic field differently, so giving the different spectral absorptions. The molecular screening constant is actually anisotropic or directional. So the magnetic interaction depends on the orientation of the molecule with respect to the magnetic field direction. As this orientation is changed by molecular movement, so the molecular motion can be followed. When the fluctuating local fields from nearby nuclei have frequency components at the resonance frequency of the nuclei of interest, they will cause relaxation.

The magnetic absorption timescale for the NMR transition is typically of the order of 10^{-7} to 10^{-9} s, and this can be relatively long compared to the time required for some molecular movements.

Again, the high resolution nuclear magnetic resonances of different conformations will have different observable frequencies. Provided that the time required for the observed resonance is sufficiently short, it is possible to observe separate absorptions from the different conformations. As with the infrared measurements, it is then possible to relate the relative intensities of the resonances to the populations of the conformational states.

However, when there are only small energy differences and low barriers to interchange between the different conformations, the exchange is so fast that only an averaged NMR resonance is observed. While the shape of the signal does change with temperature, it is difficult to use this as a measure of the conformational energy difference.

In proton NMR, it is usually difficult to observe resonance of a specific atom uncoupled from that of its neighbours. However, ^{13}C NMR can select and measure resonances that are uniquely associated with specific atoms in the polymer. In this respect, ^{13}C NMR is almost unique amongst the methods available.

Since the torsional motion usually involves movement of protons, it can also be observed in inelastic neutron scattering. However, once again, the required measurements are difficult to make, as is discussed below with reference to solid polymers.

16.3   Kinetics of conformational change

Kinetic information on the molecular conformational change can be extracted from dynamic mechanical studies, as described in Chapter 10, from the closely related acoustic relaxation experiments described in Chapter 11, and from dielectric relaxation covered in Chapter 12. In all of these, the observation of a transition in the frequency dependence of the property under study yields a relaxation time for the molecular process. This in turn transforms into the kinetics of the movement. Again, the activation energy associated with the conformational change is obtained from the effect of temperature on the relaxation time, using either the Arrhenius equation or a related analysis.

Much of what is known about the dynamics of molecular movement has come from such studies. However, these relaxation techniques have an intrinsic problem. They indicate the way in which the rate of a process changes with temperature, but, unlike resonance spectroscopic methods, do not identify the particular molecular transformations which are responsible.

Information on the reorientation of specific groups in a polymer molecule can be obtained by extending the time dependent photo processes, described in

Chapter 13, using polarised exciting light. Examination of the depolarisation of the resulting luminescence allows orientation of a specific chromophore to be selected and its rotation studied. This can be done in steady state experiments which yield the amount of rotation of the excited state during its lifetime. More informative are time dependent observations of the polarised and depolarised emissions. Unfortunately, most common polymers do not have chromophores that can easily be studied in this way. Luminescent probes can be attached to the polymer backbone and these used to monitor the motion. However, such measurements usually indicate the way in which the probe is attached to the polymer rather than the intrinsic motion of the polymer chain.

In both the spectroscopic and the relaxation measurements, the most direct time resolved information about rapid processes is obtained using pulsed excitation. This allows the collection of data from multiple pulses extending over convenient measurement times.

16.4   Studies in solutions and gels

Analyses using the techniques described above require the gross translational and rotational motion of the molecules to occur on a timescale different from that for conformational changes. This is usually the case in dilute solutions, but not necessarily in more viscous media.

The problem with concentrated solutions, gels, melts and solids is that the timescale for diffusive motions can become comparable with that for the conformational change, so it is then difficult to separate these different types of motion.

Since the dynamic behaviour of polymer molecules in solution depends critically upon concentration, it is convenient to start with dilute solutions. In very dilute solution, the polymer molecules will execute motions which are characteristic of isolated chains. This is the situation associated with the *normal modes* discussed in Chapter 8. These are the result of collective motion of large sections of the chain and cause the shear dependent viscosity of dilute solutions to be considerably greater than that of the solvent.

Interestingly, for long chains, the time required for the polymer to diffuse a distance which corresponds to its own length can be of the order of 10^{-4} to 10^{-3} s or even longer. On the other hand, conformational changes usually occur on a timescale of 10^{-5} to 10^{-8} s. So, many conformational changes may occur during the gross molecular diffusion process, as has to be the case for *reptation* to occur.

The most common way of studying the motion of polymer molecules in such media is to use relaxation techniques to measure the effect of shear rate on, or the frequency dependence of, the real and imaginary parts of the modulus, the compliance or the viscosity as discussed in Chapter 8. In so doing, it is usually assumed that changes in constraint, such as temperature, will affect only the rate at which the polymer molecule moves, and will not change the nature of the movement involved.

A number of experimental methods exist that allow polymer solutions to be subjected to different shear rates or to *oscillatory shear*. Data obtained over a given range of shear rate, or frequency, are shifted to form a universal curve (as in the use of the Williams–Landel–Ferry equation, explained in Chapter 4). This can then be compared with the predictions of various models such as those proposed by Rouse or Zimm. The former assumes that there is minimal interaction between the solvent and the polymer, and is sometimes referred to as the *free draining model*. In reality, there is some interaction between the solvent and the polymer chain. This is addressed in the *Zimm model*, where the drag introduced by the solvent influences the motion of the chains.

However, as the concentration is increased, the polymer molecules will start to interact and ultimately, if they have sufficient chain length, they will become entangled and form a gel. Radiation will be scattered by regions of different density in the solution or gel and so be used to give relevant molecular information. This can be done by using *pulsed scattering techniques*. While steady irradiation does not yield time dependent information, by using a pulsed source it is possible to observe how the scattering parameters change with time as well as wavelength. Neutrons or laser light photons, often in polarised form, can be used for this.

Such studies show that when the chain length and concentration are greater than the critical values for entanglement, reptation, normal mode motions and conformation change together give rise to relaxation features on a timescale of 10^{-3} s or slower. As the concentration is increased further, to above the gel point, translational motion ceases altogether. However, the segments of the polymer chain are still able to rotate quite freely.

At this point it is appropriate to reappraise the timescales on which various types of motion occur (Figure 16.2).

16.5   Studies in solids and melts

As indicated in Chapter 4, phenomena such as the glass to rubber transition are associated with the onset of movement in the polymer backbone. In the solid phase, the diffusion and long range motions have been stopped, and so the observed characteristics reflect the collective segmental motion of a number of backbone elements and the very local motions which only involve rotation about a single bond. These latter motions are usually thermally activated and so their relaxation times are not sensitive to pressure. In contrast, the segmental motions

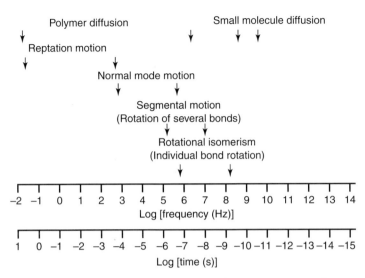

Figure 16.2 *Summary of the timescales and frequency ranges for various types of motion in solution and in the melt.*

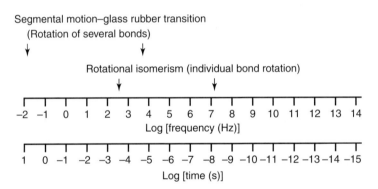

Figure 16.3 *Summary of the timescales and frequency ranges for various types of motion in a solid.*

require the availability of free volume and so the relaxation times are sensitive to pressure. Consequently, they cover a wider range of relaxation times. The frequency ranges and timescales typical for such motions in polymer solids are summarised in Figure 16.3.

The smaller number of techniques used for the study of polymer solids are broad band dielectric relaxation, dynamic mechanical thermal analysis, small angle inelastic neutron scattering and NMR, which together span a frequency range of 10^{-2} to 10^{-10} Hz. So, they can be used to study all the correlation frequencies that characterise the motions of interest.

One of the most popular methods of studying such transitions is *dynamic mechanical thermal analysis*, DMTA. The sample is subjected to some form of oscillatory distortion, as described in Chapter 10, and its response observed as a function of frequency and temperature. In all cases, the scale of the motion being observed is relatively local, usually limited to the motion of eight to ten units of the polymer. The shape of the response curve conforms to that of a simple relaxation or a limited distribution of relaxation processes. It is only when the polymer is in the melt phase that the longer range motions become apparent.

Dielectric measurements of these transitions exhibit very similar features. Of course in this case, the magnitude of the phenomenon is governed by the magnitude and movement of a dipole in the polymer.

NMR can be used to study the relaxation of polymer molecules in the solid state. Slower motion will render broader and overlapped spectra, as is generally seen in solid samples. As molecular motion becomes faster, as in solutions or in polymers much above the glass transition temperature, the screening constant will be averaged and the spectrum will become narrower, yielding a high resolution spectrum. The spectral line shape can therefore provide information on molecular motions in the samples.

If the spectrum is not so narrowed, it is necessary to orientate the sample at a so-called "magic angle" to the magnetic field and to spin it at high speed.

Pulsed sequences are used to decouple the various contributions to the nuclear magnetic interaction. Since most spectrometers operate at frequencies of the order of several hundred megahertz, the timescale of the fluctuations observed are of the order of 10^{-5} to 10^{-7} s and hence are dominated by the segmental motions of the polymer. However, using special pulse sequences, these timescales can be extended to 10^{-2} s, when the influence of normal modes becomes evident. Such pulsed measurements are usually made at a single frequency, so it is difficult to obtain a meaningful picture of a distribution of relaxation processes.

Inelastic neutron scattering can be used to study very high frequency molecular motions. This is because neutrons have wavelengths in the order of atomic or molecular dimensions and possess energy that is comparable to the thermal energies of atomic motion. The energy that may be exchanged during inelastic scattering processes can then be measured experimentally.

When neutrons collide with nuclei within atoms or molecules, inelastic scattering can occur. In the process, both the direction and the frequency of the scattered neutrons will be changed. The intensity of scattered neutrons will be recorded as a function of the scattering angle and the magnitude of the energy exchanged. The magnitude of momentum transferred determines the shape of a spectrum and generally the half-width at half-maximum of the peak increases with increasing transferred momentum. Since the positions of all nuclei in the scattering system change with time, this movement determines the shape of the scattered pattern. Three representative types of motion can be observed. These are vibration, rotation and translation, each occurring in distinct regions of the inelastic neutron scattering spectrum. The lowest scattering angles and energy reductions correspond to the polymer molecular motions that are the subject of this book.

However, theoretical modelling and simulations generally have to be used to interpret the experimental data. This means that the technique can yield parameters of an assumed process, but not verify what that process actually is.

Further reading

Collyer A. and Clegg D.W. *Rheological Measurements*, 2nd edn., Chapman and Hall, London, 1998.

Higgins J.S. and Benoit H.C., *Polymers and Neutron Scattering*, Oxford Science Publications, Oxford, 1994.

Komoroski R.A. (Ed.) *High Resolution NMR Spectroscopy of Synthetic Polymers in Bulk*, VCH Publishers, Deerfield, 1986.

Urban M.W. and Craver C.D. (Eds.) *Structure Property Relations in Polymers*, Advances in Chemistry, Series 236, American Chemical Society, Washington, 1993.

Index